The Day They Were Swept Away

10 Survivors and Their Unbelievable Stories of Being Shipwrecked, Adrift, and Stranded at Sea

Oliver Martin Cass

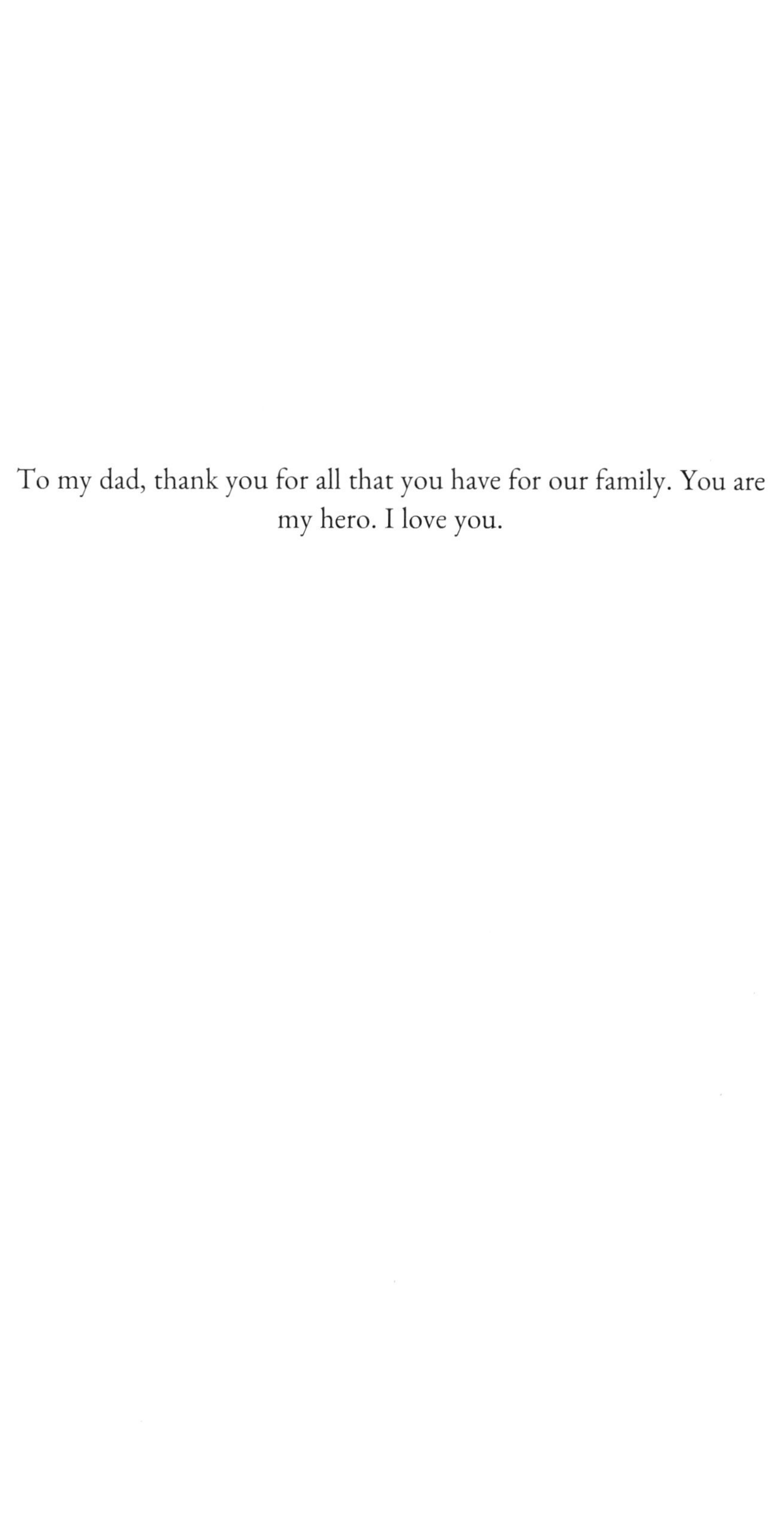

To my dad, thank you for all that you have for our family. You are my hero. I love you.

SPECIAL BONUS!

Want this Bonus Book for Free?

Get this book for free and also have access to all my new books by joining The Fanbase!

Scan with
your camera
to join!

Contents

Introduction

Humans may have had a fascination with flying for centuries but didn't act on it because there wasn't the technology to get a person into the air. This was not so when it came to seafaring. One of the essentials of life is water, so humans have always been around it. Not only did this resource quench their thirst, but it also provided food and transportation.

Seafaring History

The earliest depiction of ships was found on clay containers and tablets that dated back to 4000 B.C.E. in Egypt. Egyptians were dependent on the Nile river for everything, including the transportation of goods. The boats depicted in clay were often long and contained oars and sails. Although these boats were perfect for ferrying goods up and down the Nile, they weren't good for maneuvering out in the ocean due to their length. This hardly stopped the development of the ship as we know it today.

Around 3000 B.C.E., the Egyptians completed one of their first ocean navigated trips to Crete. They achieved this by hugging the coastline and using the landmarks they could see to navigate. Later, they would use channels emptying out into the Red Sea to travel further.

During this time, warships and cargo ships were drastically different as each had different roles they needed to fill. While the warships were built for speed and carrying fighters, the cargo ships carried goods with as few people as possible to man the ship. This resulted in the cargo ships being rounder while the warships were longer.

Designs of ships greatly differed from continent to continent, influenced not only by the people who made them but by the job they were fulfilling. Rowing was the main form of power for ships. This was until people realized that, when on open water, this method of power didn't provide as much momentum as they needed—especially when there were storms.

In the beginning, the sail was spun to catch wind from the back (known as a following wind) to boost the ship's speed. However, this often meant the ship would have to wait for not only favorable weather but also wind from the correct direction. As humans' driving nature caused them to want to explore the world around them, this unnecessary pause wouldn't do.

It was between 1200 C.E. and 1400 C.E. when humans started building ships to have multiple masts and sails. The downside was that more people were needed to hoist sails as their sizes grew. With trade and exploration fueling their resolve, the shape of the ships and sails changed to get the most out of the ship. By the 16th century, massive ships known as galleons were floating around with untold riches being transported through the then known world.

This brought danger to those on the ships, as sea life on a literal floating bank was dangerous. Pirates, privateers, and any number of enemies would take a chance at plundering these treasures. Although able to defend themselves, most galleons were unable to flee from the faster ships pursuing them. Even if they weren't being pursued, it would take months, if not years, for these ships to arrive at their intended destinations. This often resulted in sailors suffering from scurvy and other ship-bound illnesses.

It wasn't till the late 18th century that other forms of propulsion were considered for ships. This was when the steam engine powered by coal was invented. However, these engines were large and required larger ships to carry them. Then, paddle boats made their first appearance. They were a novel form of travel but were slower than a carriage on a good road. Eventually, these would fall out of transport used during the 1800s.

Oceanic steamboats were also falling out of favor as they required vast amounts of coal to be powered. Coal was better used on the railway. Not wanting to lose the ability to travel by sea for leisure, the engines were once more redesigned to be double- and even triple-expansion (reciprocating engine). This dramatically increased speed. However, when a piston broke there was a chance that the engine would be destroyed, stranding the ship.

By the 19th century, paddles were replaced with propellers and hulls were made with steel. By the end of the 19th century, the steam turbine engine was invented and was found to be faster than the reciprocating engine. First, it was powered by coal, but later it was by diesel which saw these ships traveling vast distances throughout the world's oceans.

Today, the diesel internal combustion engine is mostly used because of the increased cost of petroleum since the early 1970s. However, some ships may even have combinations of different types of engines, depending on their job. Meanwhile, some ships try to leave a smaller carbon footprint by using a gas turbine to power them.

Most modern ships are built for transporting cargo rather than ferrying people. This excludes cruise ships. These ocean liners are built to get the most stability they can with the current shipping materials available to those who build them. Cruise lines such as the Royal Caribbean International not only try to balance the fuel and cost of each trip, but also the impact they have on the environment. Ships may not have undergone as many changes as planes through the years, but they are still a mode of transportation and relaxation for many people around the world.

Is Ship Travel Safe?

Similar to airplanes, there are many regulations in place to protect the ships, passengers, and crew from possible disasters. Shipping was one of the first industries to implement international safety standards. It's regulated by many United Nations agencies, including the International Maritime Organization (IMO), which is in charge of the safety and security of the ship and pollution prevention.

Many other regulations are in place to prevent any catastrophes from occurring. The International Convention for Safety of Life at Sea (SOLAS) requires regular surveys of ships, ensuring they meet the minimum standard in safety. This covers ship construction and basic safety equipment. The International Safety Management Code is aimed at shipping companies, ensuring they have the necessary licenses to operate. There are regular audits to make sure these companies abide by this. Another important regulation is the convention on International Regulation for Preventing Collisions at Sea (COLREG), originally known as the Collision Regulations. These are pretty much the "rules of the road" when traveling on the sea. It explains who has the right of way and prevents any possible collision.

These are hardly the only regulations. There are others that govern the training and standards of the crew and even how much or what cargo a ship may carry. There are even some regulations that are meant just for either cargo ships, passenger ships, or ports. Despite all these regulations, accidents still occur.

Accidents

Thankfully, accidents involving large passenger ships during peace times are few and far between. The last major accident was in 1987 when the Philippine ferry *Doña Paz* collided with an oil tanker at night, causing it to catch fire before sinking. Over 4,000 people lost their lives in the accident. The ferry was operating without a license, was poorly maintained, and had been suspected of carrying nearly double its allowed load of passengers. Even in the last 10 years, passenger ships have still gone down. Thankfully, the number of deaths are not as high as seen with the *Doña Paz.* It isn't just registered passenger ships having accidents, either, as recreational boating still sees its fair share of unnecessary deaths every year.

According to the U.S. Coast Guard, in 2020 there were 767 recreational boating fatalities (Fredrickson, 2021). This was from 5,265 accidents which also resulted in 3,191 non-fatal injuries. This was a drastic increase from 2019 (more than 25%). The leading causes of these incidents included inattention, improper lookouts, increased speed, inexperience, and machinery failure. Alcohol consumption was responsible for about 18% of the total fatalities seen in 2020 (more than 100 deaths). Of the known deaths, 75% were due to drowning, of which 86% weren't wearing lifejackets.

A total of 77% of the accidents were caused by an operator who didn't receive boating safety instruction. Most accidents occurred on open motorboats (50%), while kayak accidents came in at 15%. The operator of the boat isn't the only one at fault. Once on a boat, everyone is responsible for their own safety. Having a sober captain and passengers, wearing a life jacket, going on safety courses, having a vessel safety check, and having an engine cut-off switch are just some of the ways you can prevent incidents. Ones that should easily be avoidable.

Despite all the regulations and safety protocols put in place, there is always a chance that something beyond anyone's control can influence the ship or boat you are on. When this occurs, there's not only a chance of death and injury, but also a chance of being left adrift at sea for an extended period of time.

Dangers

When a shipwreck occurs, you and other passengers can be injured. Some injuries can be superficial, such as cuts and bruises. Other times, people may not be as lucky. When a collision occurs, the sudden stop can cause people and objects to fly forward. This can result in spinal cord injuries, whiplash, broken bones, snapped ligaments, and traumatic brain injuries. Even when surviving the initial impact, there's a chance there could be fire and debris you have to get through to get to places of safety, such as the lifeboats. Cuts and burns may not seem like serious injuries, but they will sap your strength when you most need it.

Other times, there's a chance you can be thrown overboard during an accident. Death and amputations will occur if you land close to the propellers. If you are lucky to avoid these, you will soon realize you are surrounded by miles of water with no land in sight. Even if you know how to swim, there is a strong possibility of you drowning if you don't find something to hold onto or get to land as quickly as possible.

After Initial Survival

Your initial survival of the boating accident is but one part of your overall survival. Depending on your situation (stranded boat versus in the water) and how quickly help can reach you, being stranded in the middle of the ocean is one of the most dangerous places to be. When on an unmoving boat, you are at least floating and have access to some resources. However, when in the water, it will be considerably more difficult to survive for an extended period.

The main concerns—after injuries—is exposure, dehydration, and exhaustion. Whether from the sun via dehydration and sunburn or water temperatures below 94°F causing hypothermia, exposure to extreme temperatures will kill you quickly. In the case of the sun, you need some form of shelter over you. In the case of being in the water, you will need to find ways to keep yourself warm. This can include swimming or drawing your knees up to your chest if possible.

With little to no freshwater, you will soon find yourself dehydrating. Dehydration will kill you within 3–5 days, depending on your condition. With no food available and you battling the elements and the ocean, eventually, exhaustion will cause you to lose the battle to live or even consciousness. When you aren't floating in or on something, there is a chance you will slip under the waves, never to be seen again. If that isn't the worst, mental fatigue and animal predation is also something you also have to be concerned about.

About the Author

I am Oliver Martin Cass, and I hail from New Zealand, where I live with my wife and two children. When I am not busy exploring our backyard and admiring the endless beauty of New Zealand with my family, I like to write and share other people's personal stories of survival and triumphs. With 13 years of working in health care as a nurse under my belt, I have grown to connect with people I have worked with along the way. Witnessing their struggles, I have learned to appreciate how people survive and thrive even at their worst. My father, who was a survivor himself of a maritime disaster, is an inspiration to my journey in writing and sharing these unbelievable and inspiring stories. This sparked my interest in what sets many survivors apart from those who failed to make it out of a tragic situation.

In this book, I have gathered the stories of 10 survivors who survived lost at sea either by being adrift or marooned. What sets these individuals apart from those who died? Is it a special skill? Or perhaps luck? Or a combination of both? Do you think you could survive being surrounded by miles of water with nothing to drink?

Explore further to discover Guinness World Records holders, people who fiction heroes were based on, and those willing to do anything to survive one more day. Follow the strife they went through, what caused it, and where they are today. There will even be a bonus chapter with tips to help you improve your chances of survival if you ever find yourself adrift at sea.

Disclaimer and Trigger Warning

All the information expressed in this book comes from interviews and historical evidence of those involved in ship accidents through the ages. It is not the intent of this book to cast blame on anyone but rather to use the evidence available at the time of writing to tell these survivors' stories and explain what occurred. The topics covered within these pages will deal with catastrophic accidents, injuries, and even death. Reader discretion is advised. To the families of the deceased, the survivors, and others caught up in the accidents, you have my deepest condolences for what you were forced to go through.

Chapter 1: Alexander Selkirk: The Real Life Robinson Crusoe

One may see that solitude and retirement from the world is not such an insufferable state of life as men imagine, especially when people are fairly called or thrown into it unavoidably as this man was. –Captain Woodes Rogers

Born as Alexander Selcraig in 1676 in Lower Largo, Fife, Scotland, this young man soon became unruly and quarrelsome as he grew up. He was the seventh child of John Selcraig, a cobbler, and Euphan Mackie. He was often spoiled by his mother, who believed him to have been born lucky. She even actively encouraged him to turn to the sea as he got older. His father was against this idea as he wanted his son to help in the family tannery and shoemaking business. This often resulted in multiple arguments between father and son.

Alexander was known for his temper and quickly came to blows. This often resulted in trouble between him and the church. On August 25, 1695, this seemed to have come to a head when the church summoned him to answer for indecent behavior done on the church grounds. Although he had been caught with his pants down, he never answered these summons as he was away at sea at the time.

Even after his return, things didn't get better. On November 7, 1701, after drinking from a can filled with saltwater, his brother Andrew laughed at his mistake. Filled with rage, Alexander struck his brother with a wooden staff. Their father, John, and elder brother, also known as John, tried to step in, which resulted in Alexander taking his ire out on them. Not even Margaret Bell, young John's wife, managed to come away unscathed.

This is likely what led to Alexander changing his name from Selcraig to Selkirk, and he was officially done with being on land. He approached William Dampier and begged to go on the next buccaneer voyage to South America. It took some doing, but thanks to his skill in geography and math, William agreed to take him on as a navigator. Alexander left with William in September 1703. William was captain of the *St. George*, while Alexander boarded *Cinque Ports* under Captain Charles Pickering.

Life aboard a ship is dismal at best and often the sailors suffer from scurvy, cholera, and any number of tropical diseases. It wasn't only the men who were suffering; even the ships suffered. Often they would be used to plunder enemies—usually the Spanish and French—and they would sustain heavy damages. This needed to be fixed as soon as possible, or they'd sink. Despite this, the *Cinque Ports* continued its journey without concerns until late November in 1703, when the captain died of a fever. Thomas Stradling, Charles' lieutenant, took up the captaincy, and this is where things started to sour.

Thomas was disliked by most of the crew as he was a strict and not very competent leader. Alexander, who was now second-in-command, often butted heads with the new captain. Morale on the ship was low, and it took William coming to the ship to prevent a mutiny from breaking out. However, this didn't soothe the troubles brewing.

The Day He Was Swept Away

The *St. George* and the *Cinque Ports* would often travel together, but in May of 1704, the *Cinque Ports* left its sister ship behind to go plundering on its own. This was nearly a fatal mistake, as, after an unsuccessful attack on a fortified Spanish-run port, the ship was damaged. By September of the same year, it was taking on so much water that the sailors had to bail out the water throughout the day and night. Not only that, but they were in desperate need of fresh supplies.

Thomas brought *Cinque Ports* to Isla Más a Tierra (now known as Isla Robinson Crusoe) a month later. This uninhabited island was but one of the islands in the Juan Fernandez Archipelago, about 400 miles away from the coast of Chile. The island hosted trees, wildlife, freshwater springs, and ample vegetables planted by those who arrived before them. Although only 18.5 square miles, the island had everything the crew needed to recuperate and restock their ship.

Alexander took this time to argue with Thomas once more. He was convinced that worms were infesting the ship and they needed to spend a considerable duration of time out of the water to fix it. Thomas disagreed and wanted to set sail as soon as the repairs he ordered were completed and the ship restocked with supplies. Alexander wouldn't have any of it, and on the day of departure a month later, he refused to set sail because of the condition of the ship. He felt that it would be better to be stranded on this island than sink out to sea later. He had hoped that other sailors would agree with him, but it soon became apparent that he was the only one who thought the ship was in terrible condition.

Thomas, wanting to make an example out of Alexander, agreed and had all his belongings from the ship returned to him. Realizing the mistake he had made, Alexander then tried to beg his way back onto the ship. Thomas, now with the upper hand, wanted to show his authority to the rest of the sailors and denied Alexander.

Saving what dignity he had left, Alexander took his belongings and returned to the island as the *Cinque Ports* left. He had just been marooned.

Alone and Unafraid: How Alexander Remained Sane

Normally, a sailor would be marooned as punishment for being a thief or mutineer. This was considered a death sentence, as the places where sailors were put to land didn't always have food or freshwater. Usually, the condemned would be given a flintlock with at least one bullet. This would be used to end their suffering if the hunger or thirst got too much.

Alexander was lucky, as he had the contents of his entire sea chest and an island brimming with supplies. He felt that he wouldn't be left on this island for long, as it was, after all, a stopping point for many privateers and pirates alike to restock on freshwater and vegetables. All he had to do was wait and he'd be picked up in no time.

He continued to live on this island for four years and four months, having almost no contact with the outside world. Despite having all the tools at his disposal to survive, the first few months were the hardest. With no one to communicate with, the loneliness was eating him alive. At times he considered ending it all with his flintlock, but in the end, he didn't. He kept himself sane by singing psalms and reading from his bible, and this seemed to fortify his mental strength.

Those first few months weren't easy. Although there were able supplies, the island was frequented by gales that were strong enough to snap the strongest of the trees. Not only that, but rats infested the island. Often Alexander would awake to the vermin chewing on his clothes or feet. Filled with the want to survive, he thought back to the stories William told about other sailors who managed to survive while marooned on islands. This gave him many ideas he would go on to implement.

Using the branches of the pepper trees on the island, with grass, he constructed two dwellings. One for sleeping in and the other to cook in, possibly to avoid rats getting into his sleeping area. With shelter taken care of, the next thing he needed was food, as fresh water was readily available. The island had cultivated turnips, planted by sailors long before he arrived, as well as a collection of other plants such as watercress, pepper berries, wild plums, and cabbage palm. He was also spoiled for choice when it came to meat. The island had wild goats, lobsters (when he could catch them), sea lions, and seals. He had tried to catch fish but found without salt, his stomach often complained after he ate them.

To aid him against the loneliness he faced, he befriended and tamed some of the island cats. This gave him companionship and protection against the nightly rodent attacks. Instead of using his flintlock to end his life, he used the flint to help him create fire.

Time passed slowly for the marooned sailor, but he soon got used to the silent days. Eventually, his clothes, bedding, and shoes started disintegrating with age. Not put down by this, he fashioned new clothes with a modified nail and goat skins. He had no way to replace his shoes, so continued to walk everywhere barefoot, developing calluses as he did so. He used everything the island provided him. Once, when his knife eventually dulled and broke, he made a new one using barrel hoops he found.

During all the time he spent alone on the island, only two ships landed before he was rescued, despite his lighting fires to attract attention. Unfortunately, both of these vessels were Spanish. When he was noticed by one of the ships, he was chased, but as Alexander knew the island, he managed to hide long enough for the ship to leave. The reason for his hesitancy to be found by the Spaniards was that they would likely capture and torture him, before sending him to the gold mines as a slave.

Another ship wouldn't arrive until February 2, 1709. The English privateer, Woodes Rogers, on the *Duke* would be one to be Alexander's saving grace. As the crew of the *Duke* landed, they were surprised by a hairy, goat-clad man, who could barely speak words but was clearly happy to see them. This happiness quickly turned sour when Alexander recognized the navigator on Woodes' ship. It was none other than William who also recognized him. Disgruntled, Alexander refused to remain on the *Duke* and insisted that he would rather stay on his island.

Yet, Woodes was so impressed by the survival skills Alexander had exhibited, he insisted the man remain on the ship and become his second-in-command. William gave credit to Alexander's skills as a sailor, which only made Woodes more insistent that Alexander remain with them. He was finally rescued.

Sailor Till the End

Despite being marooned for over four years, Alexander didn't return to Scotland until October 1711, eight years after leaving. After earning £800 from plundering the *Nuestra Señora de la Encarnación y Desengaño*, he returned home for a few years. He lived off his riches but avoided companionship with others. Branded a loner, he would often be seen staring out to the ocean, longing for the island that had been his home for years.

Then the ocean bug bit him once more, and in 1717, he became a Royal Navy officer and returned to the wide-open ocean. He continued to sail until December 13, 1721, when he died of a tropical disease—likely yellow fever. He wasn't the first that year. He was buried at sea.

Although Alexander met his end, his tale is still around for all those who are curious. Perhaps you have already read this book. In 1719, Daniel Defoe published his book, *Robinson Crusoe*. The book is believed to have been based on the life story of Alexander. This book has been around for over 300 years and is still enjoyed by many people. Yet, this isn't the only book that describes what he went through. Woodes Rogers also published a book in 1712 called *A Cruising Voyage Round the World*, which is considered the earliest recordings about Alexander's extended stay on Isla Más a Tierra. It is also likely the base from which many stories inspired by Alexander were written. John Howel also published a book in 1829 called *The Life and Adventures of Alexander Selkirk*.

Each book has its own unique telling of an unruly, quarrelsome man who spent years in near-silent contemplation on an island far away from other humans. Regardless of the telling, Alexander had to overcome his own mental weakness to survive being without human contact for over four years. In the end, despite returning home or sailing the wide-open ocean, he could only think back to that island and the peace it had brought him.

Chapter 2: The Man With Two Records: Oguri Jukichi's Incredible 484-Day Ordeal at Sea

> *His survival story is very impressive. I admire his spirit. He even kept his promise to his crew to build a memorial for them.* –Katsunobu Mizuno, head of the club for Jukichi, Captain of the Sea

Not much is known about the early life of Jukichi, except he was born in 1785 in Sakushima in the Aichi Prefecture. He had no last name—as was the norm for this period in Japan—and his biological parents were unknown. He would be later adopted by a loving couple who lived in Handa, Aichi Prefecture.

During this time—the late Edo Period—Japan was a closed country, with few foreigners allowed within its borders. Despite this, it can only be guessed as to what Jukichi's upbringing must have been like. He started work early and turned to the sea for employment at 15. He clearly was a hard worker as, 14 years later, he became skipper of the freighter *Tokujo-maru*. The *Tokujo-maru* was a junk—a sailing ship with full battened sails used in China and Japan—and was usually used as a middle-class merchant ship. Its home port was in the Nagaya area, also in the Aichi Prefecture.

One can only believe that Jukichi was a good captain and treated his men fairly as they toiled, transporting goods around Japan.

The Day They Were Swept Away

On November 14, 1813, Jukichi was returning from Edo (now known as Tokyo) with a cargo of hundreds of bags of dry soybeans. The trip there had been uneventful, but as they were making their way back, a storm started to brew off the coast of Shizuoka, Shizuoka Prefecture. It wasn't long before the howling winds whipped the ocean into a frenzy as the storm struck them.

The *Tokujo-maru* was an unwieldy ship at best, better suited for the shallower waters around the coast of Japan. It was not fit to be out in deeper waters. The giant swells made short work of the rudder, making the helm useless for steering. Sensing the danger they were in, Jukichi ordered his crew to cut the mast down before the wind brought it down. At least if they brought the mast down, there would be less damage to the ship. The storm continued to batter the ship until it eventually dissipated.

With the loss of the rudder and the mast, the crew of 15 had no way to steer their ship or use the wind to power them back home. They were at the mercy of the currents of the Pacific Ocean, which was currently dragging them eastward. At this point, the entire crew was safe and likely uninjured.

Slowly Starving: Incredible Fortitude and Luck Saved Three

The only saving grace for the crew of the *Tokujo-maru* was that it had ample food stores for them, even if it was just dry soybeans. The first thing the crew did was to ration the beans and what rice they had left from their provisions. They had no idea how long they would be adrift until they could be rescued.

To get drinkable water, the crew built a homemade distillation process that used seawater to create fresh drinking water. This was likely a type of solar still. For a time, the crew survived this way, occasionally getting the opportunity to add fish to their diet when they could be caught. However, this diet wasn't good enough. When a person doesn't get the necessary vitamin C in their diet, they can start to suffer from scurvy—which is what started to happen to the crew.

Historically, long ship voyages were often plagued by scurvy, as the crew couldn't keep fresh fruit and vegetables from spoiling. The effects of the disease start to appear within 8–10 weeks when insufficient vitamin C is consumed. Vitamin C is vital to the body as it is used in many functions. This includes the absorption of iron and maintaining connective tissue (collagen), and it's also used in the creation of many enzymes the body needs to function. The first signs of scurvy are fatigue and spotty skin. As the disease progresses, teeth fall out, skin turns yellow, old wounds open up (scar tissue breaks apart) and start to bleed, wounds fail to heal, anemia, and finally, death.

Without the necessary vitamins and minerals needed to keep their energy levels up, the crew started to fade. By March of 1814, most of them had given up hope of ever being found alive, adding more stress to their already stressed bodies. Many crew members were starting to become ill.

The first death occurred in May, and by late June of the same year, all but three of the initial crew had died. This was both a curse and a blessing to those who survived to this point. They had more supplies they could survive on. The survivors were Jukichi, Otokichi, and Hanbe. It is likely these crew members who lasted the longest were those who had access to fish more regularly. There are low levels of vitamin C in fish and survivable levels in fish roe (fish eggs). Vitamin C is water-soluble and heat-sensitive. As the men couldn't cook the fish, they likely managed to get some form of vitamin C in their diet.

These three men continued to hold out hope as the year faded into 1815. Having spent more than a year lost at sea, the last two crew members had given up hope and were starting to fade. Jukichi had given up trying to boost morale. He had noticed mountains in the distance at some point, but with no way to steer or move the ship in that direction, he had been forced to watch them fade from his vision.

Then by some miracle, on March 24, 1815—after being lost at sea for over 400 days—the crew noticed sails on the horizon. One can only imagine the joy they must have felt at seeing human life after all this time.

Captain William J. Pigot was a fur trader aboard the *Forester* when his crew noticed the ship in trouble. Although the ship flew the British flag, it was American-owned. William had found the *Tokujo-maru* some 300 miles from Point Conception, the headland of modern-day California. The crew of the Tokujo-maru had been adrift for 5,400 miles. The *Forester* saved the nearly starved men and took them to land—despite the language barrier preventing accurate communication between crews.

Once the two crews had landed, Jukichi first believed they were back in Japan and was momentarily confused about where the men, who had saved him, had hailed from. At first, he believed them to be Dutch—who had traded with the Japanese before—but later learned the truth. Eventually, he was told they had landed at a place known as New Spain. Here he would remain for 10 days while he and what remained of his crew recuperated from their hellish journey.

The only reason the *Forester* was in the area was that it was overwintering in more pleasant weather before heading back to Alaska. Once the crew of the *Tokujo-maru* had recovered, they once more bordered the *Forester* which took them to Alaska. From there, they went to the Kamchatka Peninsula in Russia to overwinter until the weather eased to allow travel again.

In May of 1816, the three survivors boarded a Russian ship and traveled to Hokkaido. Unfortunately, Hanbe didn't survive the trip, and only Jukichi and Otokichi managed to return home by July of the same year. This wasn't the end of their misfortune, though. Once they were back in Japan, they were detained and questioned at length about where they had been and what happened. This was likely to avoid possible spies from infiltrating the country. Eventually, the two men were released, and they were able to return home.

Begging Forgiveness

Jukichi's social standing increased dramatically after surviving his ordeal. So much so that he was presented with not only his second name, Oguri, but he was also allowed to carry a sword openly in public. This was a huge honor and a status symbol of the time.

Yet, Oguri Jukichi (the Japanese write their last name first) didn't sit idle and waste his remaining days. As captain of his vessel, he felt obligated and guilty over the lives lost on that perilous journey. For seven years, he told tales of his adventure and showed all the memorabilia he had collected during his time traveling back to Japan. This allowed him to raise the necessary money to have a monument built to honor the brave men who died. Today, the boat-shaped monument can be found in a quiet temple near the Atsuta Shrine.

He continued to live for many years afterward, dying in 1853, presumably from old age. Although there is little information about his life other than his amazing journey, the world got to know him for two very important records. He and his surviving crew were considered the first Japanese to step foot in California. Secondly, they still hold the Guinness World Record for the longest time adrift at sea, with a total of 484 days. The only other person to have come close to this record is José Salvador Alvarenga, with 438 days adrift.

In Japan, he has a fan club to this day named Jukichi, Captain of the Sea. His adventures were documented and written in a book by Ikeda Hirochiki in 1822. It was titled *Funaosa Nikki* (*A Captain's Diary*), and many books were based on it. Eventually, in 2005, it was translated into English.

Chapter 3: Violet Jessop: The Incredible Story of the World's Most Unsinkable Woman

> *I was still clutching the baby against my hard cork lifebelt I was wearing when a woman leaped at me and grabbed the baby, and rushed off with it, it appeared that she put it down on the deck of the Titanic while she went off to fetch something, and when she came back the baby had gone. I was too frozen and numb to think it strange that this woman had not stopped to say 'thank you.'* –Violet Constance Jessop

Violet Jessop was either born under a lucky or unlucky star. Born close to Bahia Blanca, Argentina on October 2, 1887, she was the first of nine children born to Irish immigrant parents William—a sheep farmer—and Katherine Jessop. Sadly, only five of her siblings survive past infancy with her, but this wouldn't be the first time death would stalk her. As a young child, she caught tuberculosis (TB), and her parents were told that she wouldn't survive more than a few months. She surprised everyone by beating the disease and continued to enjoy her childhood.

Tragedy struck once more when her father died from complications of surgery. She was only 16 at the time. Katherine decided to move the family back to England, where she found employment as a ship stewardess with the Royal Mail Line. Unfortunately, she fell ill, and Violet was forced to give up convent school to bring in money for the family.

Finding work was no easy task, as she was only 21 and no one wanted to hire her. Usually, it was middle-aged women who took up the profession of ship stewardesses. Most hiring places cited their reason for not giving her employment as because she was too young and too beautiful. They were worried she would cause problems or at the very least be a distraction to the passengers and crew she worked with. She was able to circumvent this concern by avoiding wearing makeup and wearing clothes that made her appear drab and shabby. This gave her the effect of looking older than she really was. This ploy seemed to work as she was hired by the Royal Mail Line onboard the *Orinoco* in 1908.

She would remain with the company only a few months before she was hired to work for White Star Line onboard the *Majestic* later the same year. In 1910, she moved over to the Royal Mail Ship (RMS) *Olympic*, also with the White Star Line.

The First Time She Was Swept Away

The *Olympic* was the first of the three sister ships Violet would work on. Sister ships are ships built with nearly identical designs and are within the same class. Violet had been nervous about working with White Star Lines because of the routes they traveled, but soon settled and started enjoying her work.

On September 20, 1911, the *Olympic* left on its fifth voyage from Southampton. The start of its voyage was uneventful and everyone was enjoying the cruise. However, when the ship entered the Solent strait (between the Isle of Wight and Great Britain), it was running parallel to the His or Her Majesty's Service (HMS) *Hawke*.

The *Hawke* was considerably smaller than the *Olympic*, just over 387 feet versus over 882 feet, but had a unique function. It was a Navy cruiser, a warship that had the function to ram enemy ships to cause catastrophic damage, seeking to sink them. What the smaller boat didn't expect was the larger boat to displace so much water that it created a suction, causing the smaller boat to be pulled toward it. The captain of the *Hawke* was unable to break away from the suction and rammed into the starboard (right) side of the cruise liner with such force that it penetrated the hull by eight feet. It created two holes in the hull just below the aft deck (deck on the rear half of the ship), one of which was below the waterline. The 40-feet-long gash below the waterline caused two compartments to flood and twisted the starboard propeller. The *Hawke* was not without its own damage, as its bow was crushed.

Unavoidable Collision: *Olympic* Versus *Hawke*

Despite the damage to both ships and the *Olympic* taking on water, both were able to make it to port with their own power. Miraculously, no one was injured or killed in the collision on either ship. The only injuries were to the ships and the pride of the Navy and White Star Line.

White Star Line blamed the captain of the *Hawke* for ramming into their ship. The estimated damages to the cruise liner were $125,000, a considerable amount for 1911. The Navy hit back by saying the massive ship was at fault, as due to its bulk, it created the suction which pulled the *Hawke* into the wake of the *Olympic*. In the end, White Star Line had to foot the bill for the damages.

This turned out to be a financial disaster for them as the RMS *Titanic* was already delayed by three weeks, and now their other big earner wouldn't be bringing any money in. To lower the duration of time the *Olympic* needed to remain in dry dock to be fixed, it was decided that the starboard propeller would be taken from the *Titanic*, as it was delayed anyway. This wasn't the first time the *Titanic* would be scavenged to provide parts for its sister ship.

The original date for the *Titanic's* maiden (first) voyage was supposed to be on March 20, 1912, but the *Olympic* once more needed a propeller. The Titanic would leave late on April 10 of the same year.

Despite many people believing that the *Olympic* and several other ships were getting too large to travel safely, others didn't echo this sentiment. Instead, what they saw was a cruise liner that had taken a collision with a warship and survived. This is what perpetrated the notion of a so-called 'unsinkable' ship.

Violet was only 23 when this collision happened and disembarked with no injury. She clearly had no fear of this event occurring again, as she continued her employment with White Star Line. In April of 1912, she was asked to work on the Titanic, and at first, she wasn't sure if she wanted to. Eventually, she accepted the post, but only after her friends and family influenced her decision.

The Second Time She Was Swept Away

On April 10, 1912, Violet boarded the *Titanic*, and her duty was to take care of the needs of the VIPs. She found the splendor of this ship rivaled that of its older sister ship. The hours were long and the pay was not that great, but every evening she had a tradition. She would stand up on deck to enjoy the cool, fresh air before taking her leave for the night.

Despite the *Titanic* being dubbed "The Unsinkable Ship" and "The Ship of Dreams" Violet continued to carry her rosary—she was a devout Catholic. She believed in the power of prayer, and considering what she had already gone through, no one could blame her for keeping the rosary with her at all times.

On April 14, 1912, just before midnight, the *Titanic* hit the iconic iceberg and started to sink, despite people believing it to be unsinkable. This time, few people would get a chance to walk away from this accident.

Unsinkable Sinks: The Tragedy of the *Titanic*

The collision with the iceberg wasn't the first of the incidents which marred the *Titanic's* reputation. As it left Southampton, its giant hulk caused the same suction as the *Olympic* had. This resulted in another ship, the *New York*, nearly being pulled into it. Despite this little mishap, the ship was soon on its way to New York City. The ship was carrying approximately 2,200 people, of which about 1,300 were passengers with tickets for first, second, and third class.

Throughout the trip, the crew had been warned about icebergs in the area and for the most part, those at the radio delivered the messages to the bridge. Despite the warnings, Captain Edward J. Smith (who captained the *Olympic* when Violet was aboard it) only slightly altered the course but didn't lower their travel speed from 22 knots. On April 14, 1912, both the *Mesaba* and the *Californian* warned the Titanic of an icefield, the latter having been stopped by it. Despite the warnings, the message from the *Mesaba* was never sent to the bridge, while the *Californian* was actively scolded by one of the wireless radio operators for interrupting him as he was working with passenger messages.

In the dark, those in the crow's nest were unable to see any disturbance in the water from icebergs because of how calm the ocean was. This was hampered even more as the crow's nest binoculars had gone missing at some point during the trip. By the time the iceberg was sighted at 11:40 p.m. and the bridge warned, it was too late.

The first officer ordered a hard-a-starboard—turning the ship to port—and had the engines reversed. Unfortunately, as the ship was too close to the iceberg, the lowering of the speed caused the turn to be too slow. This resulted in the starboard side scraping the iceberg. Five of the so-called 'watertight' compartments toward the front of the ship ruptured and started to flood.

Knowing they were in danger now, the first distress signal was sent at 12:20 a.m. on April 15. Sadly, most vessels were either too far away (*Olympic*), had their radio turned off (*Californian*), or needed time to get there (RMS *Carpathia*). The *Carpathia* was at least 58 nautical miles (66.5 miles) away and would need at least three hours to get to them.

Violet could hardly believe she was once more on a ship that had crashed. She couldn't panic as she had a duty to the passengers on the ship. On the upper deck, lifeboats were being prepared before women and children were ordered to get into them. Violet continued doing what she could to help passengers until she was told to go to the deck. Once there, she was instructed to put on a lifebelt and show the non-English speaking passengers how to do it, as they were understandably terrified. She complied and continued to help those who were ushered onto the limited number of lifeboats (20) that were being lowered into the water.

The *Titanic* had more than the required lifeboats according to the British Board of Trade. The problem was that each could only carry 65 people at capacity, nowhere near the number of people on the ship. What made this worse is that the crew who loaded the lifeboats were not loading them to their capacity. They were concerned that the davits—cranes used to lower the lifeboats—wouldn't bear the weight. However, these davits had been tested in Belfast, but the crew was unaware of this, as the lifeboat drill meant earlier on the 14th had been canceled. This led to them being filled and placed in the water haphazardly.

Eventually, Violet was instructed to get onto lifeboat 16 to show people it was safe to do so. She even had a baby thrust into her arms, who she continued to look after for over eight hours before being found and rescued by the *Carpathia* from the Cunard Line. Freezing and numb from the shock, she barely reacted when the baby was pulled from her arms by an unidentified woman, who quickly disappeared. Only 705 people survived the sinking of the Unsinkable Ship.

Anyone who had survived the ship breaking in half and sinking found themselves in 27°F water. Most of those in the water quickly succumbed to exposure before any lifeboats turned back to help them. It was believed that the lifeboats delayed returning to the accident as the survivors were concerned over those in the water swarming the boats and toppling them.

Violet was only 24 when she survived her second ship accident, once more completely uninjured. Even after over 1,500 people lost their lives, she still had the mental fortitude to continue with this line of work. The reason behind the high number of deaths wasn't just a few lifeboats. The captain failed to sound the general alarm. This caused most of the crew and third-class passengers not to realize the danger they were in. There was too much confusion over what was happening. A few of the deaths were attributed to women who refused to leave their husbands and grown sons behind. A total of 710 third-class passengers died, as their quarters were on the lower decks of the ship. The lower decks were notoriously difficult to navigate to get to the top deck.

This tragedy changed many maritime rules—namely that radios were now manned 24 hours. Other major changes included the number of lifeboats that had to accommodate every person on the ship, passenger and crew alike. There was also the birth of the International Ice Patrol. It became their job to warn ships within the North Atlantic shipping routes of any icebergs, while they also assisted in breaking up the ice.

The Third Time She Was Swept Away

The first world war was declared on July 28, 1914. This caused many cruise liners to be drafted into service of the Navy. The *Britannic* was pushed into service the first time in 1915 and then later again in 1916. This was also true for Violet, as she found herself a nurse for the British Red Cross aboard White Star Line's His or Her Majesty's Hospital Ship (HMHS) *Britannic*. This was another sister ship to the *Olympic* and *Titanic*. Now 29, Violet was onboard a ship that operated in the Aegean Sea, transporting British soldiers from the Mediterranean and Great Britain. Despite the horrors of the Great War, she continued to serve at her post well, helping injured soldiers.

She likely would have been happy with completing her duties until the war ended, if it wasn't for one more shipping accident. This one left her injured for the first time.

Big Blast: The Sinking of the *Britannic*

The peace was destroyed by an explosion that rocked the ship on November 21, 1916, at 8:15 a.m. At first, no one was sure if they had been torpedoed or if they hit a mine. Either or, the ship was going down fast. They were just off the coast of the Greek island Kéa when the event occurred.

The explosion had caused massive damage as five of the 17 watertight compartments started to flood. For the most part, the ship managed to remain afloat, but Captain Charles Alfred Bartlett started the evacuation process when it started to list toward the starboard side. As the lifeboats were being prepared, the engine was still running. However, the crew was worried that the listing would make launching the lifeboats more difficult. They then proceeded to lower two into the water. This was done without orders and before the engines were killed.

The first thing that ran through Violet's mind was that she needed to get her toothbrush. Violet had lost hers, while on the *Titanic*, and it had stuck with her these last four years. Violet boarded one of the lifeboats clutching her toothbrush, hoping that this wouldn't be a repeat of what she had experienced four years ago. In a way, she was right, there weren't as many deaths, but this event would leave her injured.

As the *Britannic* was going down, its propellers were partially above the water. The churning it created was pulling lifeboats toward it. At the speed they were spinning, no one would escape unharmed. Fearing for her life, Violet jumped into the water, only to be sucked in under the ship. Her head struck the keel—the backbone of the ship that runs along the central line at the bottom of the hull, stretching along the ship's length—under the water before coming out the other end. Luckily, some survivors pulled her into their lifeboat and she wasn't more seriously injured.

The propellers ended up sucking in two lifeboats on the port side, killing all within, before the engines were shut down. The evacuation continued until the captain noticed the ship was taking on water slower than he thought. He was convinced he could reach Kéa if he turned the engines back on. He stopped the evacuation and ordered full speed to the island. Unfortunately, with the list, the open portholes on the starboard side allowed more water on board. Soon it was apparent that the ship was taking on far more water than expected. The captain killed the engines and continued the evacuation process until the ship was clear of everyone. Within 55 minutes, the Britannic was completely submerged.

Violet attributes her survival to her thick auburn hair, which likely cushioned the blow she took. Since that day, she had been plagued with headaches. It was only when she went to see a doctor several years later that it was discovered she had actually fractured her skull against the keel. Only 30 lives were lost when the ship sank. The likely reasons for fewer lives lost were the water temperature was higher, there were more lifeboats, and there were many fishermen in the area who could give assistance.

Sailing Till Retirement

Violet clearly had had enough at this point, and she gave up sailing with White Star Line after the *Britannic* sank. She remained on dry land for four years before returning to the company to continue her occupation. She would eventually switch over to Red Star Line and traveled on the *Belgenland* for two world cruises. After some time, she returned to the first company, Royal Mail Line, and worked with them until she retired at 61. She had had no further incidents during this time.

She even managed to find love, for a time. Although she had been proposed to three times while working, she ended up marrying another steward. John James Lewis and Violet were married on October 29, 1923. Sadly, this union didn't even last a year before she divorced him and carried on with her career.

Violet retired to Great Ashfield, Suffolk to a quaint cottage with a thatched roof. She was content to raise chickens and tend to her garden. Her home was decorated with items she had managed to gather during her long career out to sea. She thought the sea was done with her, but it wasn't quite yet.

One day she received a phone call asking if she had saved a baby from the *Titanic*. When she confirmed this, the caller stated that they were that baby before laughing and putting the phone down. Many people believed this to have been a prank orchestrated by the children of the nearby village. However, Violet was adamant that she had told no one about the child she had rescued and that it truly was the child who had phoned her as an adult. She was affectionately known as Miss Unsinkable.

During her retirement, Violet had time to jot down her memoir—which was filled with pseudonyms. This book titled *Titanic*

Survivor was only published in 1997, well after her death at 83 from congestive heart failure on May 5, 1971.

Depictions of Violet are found throughout the media and books. *A Night to Remember*, published in 1955 by Walter Lord, was adapted into a film by the same name in 1958. Directed by Roy Ward Baker, it is believed the stewardess played by Etain O'Dell was likely based on Violet. In the 1997 film *Titanic* directed by James Cameron, a stewardess by the name of Lucy is seen showing people how to put on a life jacket. It is believed this role was based on either Lucy Violet Snape—who died in the sinking of the Titanic—or Violet. There was even a stage play called *Belgenland* which had a role named and dedicated to what Violet had done, unlike many other representations.

Chapter 4: The Man Marooned on His Own Ship: Sir Ernest Shackleton and the Story of the Endurance

I had a dream when I was 22 that someday I would go to the region of ice and snow and go on and on till I came to one of the poles of the earth. –Ernest Shackleton

The tale of Sir Ernest Shackleton is one of improvisation, adaptation, and survival. He was born on February 15, 1874, in County Kildare, Ireland. He was the second child out of ten but the first son of Henry Shackleton and Henrietta Letitia Sophia Gavan. When he was young, his parents moved their family to London. Ernest was often bored by what he was taught in school, leading him to not reach his full potential. His father wanted him to follow in his footsteps and become a doctor, but Ernest had a different idea. The sea was calling to him.

Eventually, his father relented and allowed his son out to sea when he was 16. By the time he was 18, he was in the Royal Navy as a First Mate, and at 24, he was a Master Mariner. This meant he was a licensed mariner with the highest seafarer qualification available. Yet, this wasn't where his dreams stopped. He wanted to reach the South Pole before any other person. He was given his first opportunity in 1901 when he and Robert Falcon Scott set out to the South Pole on the Royal Research Ship (RRS) *Discovery*. Together they were the first humans to get as close to the South Pole as possible without reaching it. Sadly, Ernest took ill and couldn't see his dream through. He was forced to return to the ship and then to England to recover.

It was likely he suffered from scurvy during this time. Because of what he had suffered, he thought of ways to combat the disease for his next voyage. He was convinced that whole foods and foods that required little cooking were the best way to combat scurvy. He was able to test this theory later when he was on the HMS *Endurance*.

Despite not reaching his dream the first time, he tried once more in 1907 on the Nimrod expedition. This time he managed to get within 97 miles of the pole before retreating due to the difficult journey. Even though he saw this as a failure, he was able to climb Mount Erebus, a massive volcano, and found the magnetic south. During the trip, the crew was forced to survive on half rations. Frank Wild wasn't fairing well and likely would have succumbed had Ernest not given him his share of rations. This act of selflessness bought the loyalty of those who had followed him. Once he returned to England, he was showered with accolades for what he had done. This included being knighted and receiving a gold medal from the Royal Geographical Society.

Before he could realize his dream, it was achieved by Norwegian explorer Roald Amundsen. This didn't stop Ernest though, as he set his eyes on a new goal. He was going to cross the Antarctic through the South Pole. The only way to achieve this goal was to use two ships: The *Endurance* and the Steam Yacht (SY) *Aurora*. The *Endurance* would have to leave from the island of South Georgia and the *Aurora* would leave from Australia.

His idea was that the crew of the *Aurora* would reach their end of the Antarctic and start setting up supply depots for the team from the *Endurance*. This would allow them to travel through the Antarctic without having to carry all of their supplies with them.

The *Endurance* left England on August 1, 1914, heading to South Georgia, to prepare for the trip of a lifetime. The crew was 27 men of Ernest's choosing and a single stowaway, which later became the ship's steward. With these men were 69 sled dogs—which would help pull them across the ice—and a single cat.

The Day They Were Swept Away

On December 5, 1914, the *Endurance* left South Georgia and made its way to the Antarctic. Within two days, the ship was already plagued by the pack ice surrounding the Antarctic. The crew bravely fought against the ice floes until January 18, 1915. A gale from the north picked up and forced the ice pack into the land. This caused the ice floes surrounding the ship to become packed tightly around it, allowing it to freeze the ship in place. They had been no more than a day from making landfall, but now they were effectively marooned on their own ship. To make matters worse, the ice started to push the *Endurance* back toward the north, regardless of how hard the crew worked at chipping away at the ice during February 1915. Ernest's dream was slowly fading away over the horizon.

The ship's radio system was too primitive and they couldn't call for help. The crew had no choice but to wait out the coming winter. During July, the pack ice was starting to break apart, but then disaster struck. The storms that arrived during August and September caused the ice to pound against the *Endurance's* hull, weakening it.

On October 24, 1915, a piece of ice finally punctured the hull and the ship started taking on water. On October 27, the pressure from the ice became so much that the stern lifted from the water. Soon after, the keel and rudder were torn from the ship. There was no way to salvage the ship, so Ernest ordered his men to remove any carriable supplies they could while they still had time to do so. It took a month for the *Endurance* to finally sink into the depths.

Fortunately, the pack ice was sturdy enough for the crew to stay and keep their supplies on. Sadly, to save on resources, the younger, weaker dogs, as well as the ship's cat, were euthanized. Frank Hurley, an Australian photographer and adventurer, documented the ship's final moments. He snapped several photos, and these would be the last images of the *Endurance* until 2022.

With three 20-feet lifeboats and all the supplies they could salvage, Ernest marched his crew over the ice. He hoped that at some point the ice would open up and they could be put to sea. He was hoping to reach Paulet Island, some 250 miles away. However, during December, the ice started to soften, and traveling over it became a difficult and dangerous task. Instead, they remained stationary on the ice until an opening appeared in early April 1916. With supplies running short, the men had been forced to eat seals and penguins, and finally, their dogs. As fresh seals and penguins contain vitamin C in their oil, scurvy was kept at bay.

Once in the water, Captain Frank Worsley managed to navigate through the rough sea from April 9 to April 15, 1916, to reach Elephant Island. It was the first time in 497 days that the crew had solid ground under their feet. Yet this wasn't a cause for celebration. The barren Elephant Island was far outside the normally used shipping lanes, and there was no hope of the crew being rescued. If they stayed here, they were going to die.

Marooned Twice: How Shackleton Saved the Crew of the Endurance

Now effectively marooned for a second time, Ernest had to think of a way to get his crew to safety. Their only hope was to travel back to South Georgia, over 800 miles away, to get help from the whaling station there. However, at this stage, the three lifeboats were worse for wear. There was no way they would carry all 28 men back. Taking the lifeboat with the least amount of damage, it was quickly dubbed the *James Caird*—who had been the chief financial sponsor—and the other boats were scavenged to make it seaworthy.

By April 24, 1916, after some recuperation, Ernest with five of his crew members boarded the *James Caird* and battled their way to South Georgia. All they had was their lifeboat, a month's worth of supplies, and each other to rely on. If they failed, the entire crew of the *Endurance* would meet a grisly fate.

After battling the ocean, they arrived in South Georgia on May 10, 1916, 16 days after they left Elephant Island. The only problem now was that they had landed on the southern shore of the island and the whaling station was on the northern part of the island. They were left with two options: get back into their lifeboat and travel by sea, or make the trek across the uncharted interior of the island.

Ernest decided against traveling by sea again. Instead, he, Frank, and Tom Crean set off to the whaling station on foot. The remaining men would wait at the lifeboat until they returned with help. The trio marched through the wilderness for at least 36 hours before reaching the station. Exhausted, hungry, but filled with hope, they had reached Stromness. They were quick to explain who they were and they desperately needed aid.

The whalers jumped into action and quickly went out to retrieve the three men stranded on the southern shore of the island. Unfortunately, they couldn't reach the 22 men still trapped on Elephant Island. Ernest wasn't willing to give up on his men.

For those left behind on Elephant Island, it would be 128 days after the *James Caird* left that they would see other human beings. It took three attempts to reach them. The first two rescue attempts were marred by ice blocking the ships from reaching the stranded men. However, the third time was the charm, and on August 30, 1916, the entire crew of the *Endurance* was safe and sound.

They had managed to survive for as long as they had by turning the remains of the two lifeboats into a shelter and hunted seals when possible. They had almost no supplies left. The hunger had gotten so bad toward the end of their second marooning that the bulk of the crew started talking about cannibalizing the first man to die, in desperation. Luckily, this never occurred.

Due to no communication during their time stranded, the crew of the Endurance had no idea what had happened to the *Aurora*. Unfortunately, they had been less lucky than their companions.

The *Aurora* had reached its intended destination and started setting up the depots as instructed by Ernest. While at Cape Evans, a gale caused the ship to break away from where it was moored. This left 10 men stranded with few supplies and an unforgiving environment. Sadly, they would remain stranded from May 7, 1915, to January 10, 1917, and lose three men. One is thought to have died of scurvy, while another two are thought to have fallen through the ice and were never found. It was after their rescue that they learned Ernest had failed in his mission. It had all been for naught.

Dreams Dashed

Despite not achieving what he set out to do, Ernest wasn't going to just fade into obscurity. His obsession with the Antarctic was one he would follow until his dying breath. Ernest returned to England in 1917 and heard about the Great War. He immediately wanted to head for the front lines. However, due to his poor health, he was forced into the role of a diplomat in South America. Afterward, he became an advisor to the Russian troops, aiding them with the knowledge of cold weather survival.

In 1919, he published *South*, a book detailing what he and the rest of the crew of the *Endurance* went through during their double marooning. While he was alive, and well afterward, many people published works about him and his famous ship.

Soon after this book, he had one more goal he set out to do. He wanted to circumnavigate the Antarctic. He wanted to map the coastline, some 2,000 miles worth, and look for any usable mineral resources on the islands scattered around it. On September 17, 1921, aboard the *Quest* RYS—the suffix for Royal Yacht Squadron was given to the ship as Ernest was a member—he set out from London to achieve this momentous task.

Tragically, on January 5, 1922, while resting in his bunk, Ernest suffered a heart attack and died. He was only 47. The *Quest* had only passed South Georgia a day earlier. The doctor on board stated that Ernest's heart attack was likely brought on by debility. Over the last few years, Ernest's health had been in decline. His body was transported back to South Georgia, where he is still buried today.

On March 9, 2022, 10,000 feet deep in the Weddell Sea, the *Endurance* was rediscovered since it was fatefully sunk more than 100 years previously. It was only four miles from where it was believed to have been sunk. The wreck is still in excellent condition considering the time it has been underwater. For now, the wreck is being preserved as much as possible and limited activity is planned to explore it further. Ernest and the *Endurance's* legacy has managed to stand the test of time, showing us that humans can endure anything when they put their minds to survival.

Chapter 5: War at Sea: How Poon Lim Survived Being Torpedoed and Adrift for 133 Days

> *I hope no one will ever have to break that record.* –Poon Lim, after being told he was the record holder of longest time adrift alone

Lim was born on March 8, 1918, on Hainan Island, China. He lived a modest life with his family and managed to go to school thanks to the money his brothers sent for his education. His life likely would have turned out as what was envisioned for him by his parents, if it wasn't for the Japanese trying to invade China. When Lim was 16, his father was concerned that his son would be drafted into the military to fight the invaders. Seeing no other way to save his son from this fate, he sent the boy to one of his brothers working on a British passenger ship.

Lim had liked the idea of sailing and agreed with his father. Unfortunately, what he didn't know was how poorly he would be treated on the ship. At the time, the Chinese on these ships were often given the worst job, the least pay, and were shoved into overcrowded, small living quarters. He became a cabin boy and worked hard, despite often being sick and teased by the rest of the crew. Eventually, in 1937, this treatment became so terrible that he swore off ever stepping foot on a ship again. Once off the ship, he moved away to Hong Kong to pursue an education in engineering.

Then World War II started in 1939. The British took heavy losses out at sea in terms of ships and sailors. They were desperate to get more people aboard their ships to aid them. They eventually asked China to aid them in 1941. Lim, still remembering the terrible treatment he had received years ago, balked at the idea of rejoining a British ship—even if the British had improved the conditions and salaries offered.

His family eventually convinced him to try once more, wanting to keep him out of the growing conflict with Japan. Eventually, Lim agreed after his cousin convinced him that the conditions aboard the British ships had truly improved for all Chinese.

On November 10, 1942, Lim was granted the rank of second mess steward as he boarded the Single-Screw Steamship (SS) *Benlomond* (sometimes spelled *Ben Lomond*) in Cape Town, South Africa. This merchant vessel would be traveling to Paramaribo, Dutch Guiana (now known as Suriname). The *Benlomond* was a tramp steamer, meaning that it had no fixed schedule and generally traveled without an escort. This wasn't a problem, as the ship was armed, but it was slow. Yet, where it was traveling, few problems were expected. However, all that changed on November 23, 1942, when a German U-boat U172 spotted them 750 miles away from the Amazon.

The Day He Was Swept Away

The U172 spotted the *Benlomond*, and under the orders of Carl Emmermann, fired two torpedoes at the hulking ship. Both tore into the ship, resulting in several boilers in the engine room exploding. The ship was going down fast, and it foundered within minutes.

Lim, being a poor swimmer at best, managed to grab a life jacket and jump into the water before the *Benlomond* sank beneath the waves. If the U-boat noticed Lim, it gave no indication of this as it slipped back into the depths of the South Atlantic. Of the 54 souls on the ship, Lim seemed to be alone in the water.

Paddling around, trying to remain out of the debris field, Lim swam for about two hours before he happened upon an eight-feet squared wooden life raft. He scrambled aboard and quickly did a recon on its contents. It had a canvas roof and a bag of a few supplies. Within the bag, he found some flares, two smoke pots, tins of biscuits, some chocolate, some sugar cubes, a torch, and a jug containing 10.5 gallons of fresh water.

Lim had no idea how long he would have to wait for rescue, so he started to ration the supplies available to him. He figured he had at least 30 days worth of rations if he was careful. His limited food and water stores were the least of his concerns. He was terrified of falling out of the life raft when the sea was rough, so he ensured he was always tied to the boat with a hemp rope. With another piece of rope, he added knots to represent the number of days he had been adrift. He soon realized he didn't have enough rope for this task and started counting the full moon instead.

He then realized his meager supplies weren't going to last long. He adapted what he had into what he needed. He fashioned a small hook from a wire inside the torch to catch fish. When this small hook wasn't good enough to land him the larger fish, he pried a nail from the life raft and made a larger hook. Any fish he caught was quickly gutted and hung from the canvas to dry. To gut the fish, he had created a knife from one of the tins. He was slowly starting to stockpile extra food. To increase his stock of fresh water, he used the canvas and his life jacket to capture any rain that fell during this time.

To improve his swimming skills, on calm days, he would lower himself into the water and swim two laps around the lifeboat. This helped him maintain his muscles, instead of letting them waste away by remaining stationary. Just when things were starting to improve for Lim, disaster struck. One day a terrible storm blew over him, and though he was able to remain in the lifeboat, his supplies were washed away or destroyed. He now had no fresh water and nothing to eat.

Battling Sharks: How Lim Fought For Survival

Within days, he was almost mad with dehydration. There had been no rain after the last storm. In desperation, Lim fashioned a bird's nest with seaweed and tried to encourage seagulls to land on his boat with a piece of rotten fish. His plan eventually succeeded. He caught the unprepared bird, snapped its neck, and drank its blood to quench his thirst. It worked for a time, but soon he was thirsty again. He needed to come up with a plan or he would die out here.

This wasn't the only problem he was facing at the moment. As he had cleaned and gutted the bird, the remains had attracted sharks to his lifeboat. Now he could no longer enter the water for his swim. However, Lim had been born on Hainan, and they were known for a particular delicacy: shark fin. He knew the large, toothed fish was edible if only he could catch one.

Taking some of the seagull meat he had been drying, he baited one of his nail hooks. He then prepared a braided line from the hemp rope available to him. With a strengthened line and a baited hook, he needed to entice a shark to take the bait. Wrapping his hands in some canvas, he dropped the line into the water and waited. He wasn't disappointed as a three-foot long shark took the bait. Lim struggled with the large fish not only in the water, but also when he managed to pull it into his small lifeboat. It took several hits with the half-filled jug over the shark's head before it was subdued.

Once the shark was no longer moving, he used his tin knife to slice its belly open and remove the liver. He drank the blood directly from it before butchering the shark to have something to eat. His prizes, the fins, were placed on the canvas to dry for a later meal.

Now that he had some supplies again, he felt a little more comfortable, but he wasn't out of danger yet. He had no idea where he was and the lifeboat had no way to navigate the ocean. He was at the mercy of the currents and anyone who passed him. Although this didn't happen frequently, it had happened on three separate occasions.

Lim had spotted a German U-boat that was practicing drills. They had been firing on seagulls when he had noticed them. Fortunately, they didn't turn their weapons on him, but sunk beneath the waves, never to be seen again. Another time, an unnamed freighter sailed right by him, not acknowledging his pleas for help. Although cold, it was an understandable move on the captain's part. Some U-boat captains were known to use a floating survivor to bait in a rescue. Once the ship started rescuing the survivor, it would then take the opportunity to torpedo the unmoving ship. No captain could risk their ship becoming a sitting duck. Lim also felt they didn't want to help him because he looked Asian.

The last encounter could have been his saving grace. A US Navy Patrol plane had spotted him and dropped a marker buoy close to where he was. With this buoy, there was a chance a ship would arrive to help him. His hopes were once more dashed when a storm pushed him away from the marker. He wouldn't be rescued this way.

During this time he often suffered from sunburn and seasickness, yet managed to keep himself sane by completing all the necessary tasks for survival. One day, he noticed the dark ocean was starting to lighten, an indication of it getting shallower. Desperate to find land, Lim paddled the lifeboat toward the lighter part of the ocean, hoping that another freak storm wouldn't send him out to sea once more.

While trying to find land, he was finally spotted by another boat on April 5, 1943. This time it was a family shipping vessel with three Brazilian fishermen. He removed his shirt and started waving it around while jumping up and down in his vessel. Noticing his distress, the fisherman went back for him. Although there was a language barrier, the fisherman knew Lim was in trouble and took him back to dry land while giving him food and water.

Three days later, they made landfall in Belem, Brazil. Due to his swimming, he managed to walk off the boat without assistance unlike many other survivors who had been adrift at sea. Lim was taken to the hospital where he would stay for four weeks to recover. He was in surprisingly good condition, only having lost about 20 pounds and suffering from dehydration. He had been adrift for 133 days and was confirmed the only survivor of the *Benlomond.* His incredible survival story made him famous among the civilian population as well as the military.

Recognized: A New Life in America

It wasn't until later in October 1943 that Lim managed to get to London. He was to be awarded the British Empire Medal by King George VI. This allowed him to use the post-nominal B.E.M. with his name. This reward was given to those of both military and civilian standing for service rendered to the British Empire. The survival skills he had exhibited during those 133 days were also incorporated and implemented in the training manuals used by the Royal Navy sailors.

In 1944, he settled in America but continued to travel the world to tell people about his amazing feat of survival. Although he had emigrated to America, the quota of Chinese citizens was filled, and he wasn't allowed citizenship. However, five years later, with the help of

Senator Warren Grant Magnuson, he got a special dispensation and was officially recognized as an American citizen.

Lim would live the rest of his life quietly in America before eventually passing away on January 4, 1991, in Brooklyn, New York. According to the Guinness World Records, Lim is credited with the longest time adrift alone in a lifeboat. Although José Salvador Alvarenga is known for his drift of over 400 days—mostly by himself—this feat has yet to be officially recognized by the Guinness World Records. Until such time, Lim is the man with the longest official record for being adrift at sea alone.

Chapter 6: Into the Storm: How Tami Oldham Ashcraft Survived a Hurricane at Sea

Afraid? I've seen the worst! I'm not afraid of this. This is nothing compared to what I just experienced. –Tami Oldham Ashcraft

Tami was practically born to be on the ocean. She had been learning how to sail from a young age while on her father's Hobie Cat (small sailing catamaran). During this time, she learned valuable lessons and survival skills. Born on February 20, 1960, in San Diego, she was only 19 when she completed her first Pacific crossing in 1979. She would complete this achievement twice. She was as comfortable on the water as she was on land, and soon she found someone to share that passion with her.

Richard Sharp, a British sailor, was every bit as experienced as Tami when they first met. They spent six months together, sailing around the South Pacific islands in her 36-foot long sailboat. Soon they were engaged, and they couldn't be happier. They both had a passion they shared and the whole world was open to them.

Then, in September of 1983, they were approached by a couple who wanted them to deliver their 44-foot long yacht from Tahiti to San Diego. The boat in question was a Trintella 44 sailing yacht registered as *Hazana*. The trip would be 4,000 miles across open water, and though a routine trip, it was the longest the couple would have ever sailed. They were confident in their skills and jumped at the opportunity to take on the challenge. The trip would take about a month if nothing were to go wrong. The couple set out on September 22, 1983.

The first two weeks of the trip went off without a hitch, but soon they heard about a tropical depression starting in Panama. This was the start of a tropical cyclone (hurricane) with maximum sustained wind speeds averaging 38 miles per hour or 33 knots. They had just passed the Equator and knew the system was heading westerly, as they had been warned about it. The storm was steadily growing in power, but there was a chance that they could avoid it if they continued toward the north.

Normally, tropical storms from Panama would peter out around Baja, Mexico. However, the warmer ocean waters caused by El Niño only seemed to strengthen the growing storm. Soon it was classed as a Category 4 hurricane. Unfortunately, about three weeks into their trip, after trying to outrun Hurricane Raymond for three days, they were caught in the worst of it on October 12, 1983.

The Day She Was Swept Away

According to the Saffir-Simpson scale, a Category 4 hurricane can cause catastrophic damage. On land, most houses would lose their roofs, windows would be blown out, and trees and powerlines would be flattened. Flying debris would cause injury or even death. Habitable areas would become inhabitable for weeks if not months because of the damage. At least on land, you can escape the destruction by evacuation; while on the open sea, there is nowhere to run.

Around Tami and Richard, the waves were over 40 feet, and the wind howled at over 140 knots (160 miles per hour). The boat, which had seemed so large, was now tiny in comparison to the waves slamming into it. They were understandably terrified, but they didn't freeze. Everything that could be battened (tied) down was tied down. Anything that couldn't was removed from the deck and stored away. The pair had some time to prepare before the worst of the storm was upon them.

The engine was still running, but the sails weren't flying. Richard was at the helm, doing his best to keep the boat upright and stay the course. There would be times that the *Hazana* would crest a wave and become momentarily airborne before being slammed back down into the water. Tami was terrified that the shuddering the ship made would cause a porthole to break open, allowing water to rush into the cabin below, sealing their fate. The wind was so strong that it ripped the anemometer—a tool used to measure wind speed and direction—from its resting place.

Richard, who was not wearing a life vest but a safety harness, was securely attached to the boat as he struggled to keep it upright. They just needed to weather the storm, and then they were home free. It was difficult to judge what the waves would do. Most of the time there was a general pattern, but occasionally a rogue wave would come from a different direction. This would cause pockets where waves would crash similarly to what they would do against the shore, but against the ship. It was exhausting work to maintain the boat.

Eventually, Richard sent Tami down into the cabin to get some rest. Tami complied, making sure her life vest and safety harness were firmly in place.

"Oh my God!" These were the last words she ever heard Richard scream. The next thing she knew the boat rolled end-over-end, like a cartwheel. The momentum slammed her into the cabin's wall, and she lost consciousness. It is likely Richard saw a massive wave and was unable to control the yacht because of it, causing it to pitchpole.

Gripping Grief: Overcoming Loneliness to Survive

Tami wouldn't regain consciousness for nearly 27 hours. She awoke partially submerged in the cabin, confused and dazed. She had hit her head hard enough to give her a concussion and split the skin in her hairline. She also had a nasty gash in her leg, which kept bumping into things as she tried to find her feet. Her head still reeling from what happened, she suddenly recalled Richard's warning he had shouted.

She rushed to the deck to see if he was injured. She was met with his safety tether hanging off the side of the boat with no Richard in sight. His safety tether had withstood the storm, but it would seem the D-ring, which kept his harness attached to the tether, had parted. Richard had been lost to the storm, with no life vest.

Overcome with grief, she called out to him a few times before realizing it was useless. For a time she was near catatonic with grief, but snapped out of it as she needed to think of her survival now. The *Hazana* was equipped with many electronic devices which could help her. Unfortunately, many of them had failed due to the damage caused by the storm. The single-sideband radio—used for long-distance communication—had stopped working two weeks into their trip and was useless even before the storm. Although she tried to get the emergency position indicating radio beacon (EPIRB) to work, it never seemed to. She opened it and found the inside had been completely corroded. It was useless. Even the satellite navigation system had been destroyed.

The only electronic device that seemed to work was the short wave radio (VHF). She tried several times an hour to send a mayday signal to anyone close by. After five days, the radio gave out, likely due to water damage. No one ever heard her mayday call.

In between making mayday calls, she inspected the ship. Somehow it was still floating, but the masts had snapped, and the torn sails were trailing in the water. She needed to come up with a plan, or she was going to die out in the middle of nowhere.

First, she needed to determine where she was. Although the satellite navigation system was broken, she could still navigate. She had a sextant, an instrument often used by sailors for centuries to track celestial bodies—sun, moon, stars—across the sky to determine the longitude and latitude. By knowing where these lines are, you could determine your rough location. Except there was a problem. It wasn't just the sextant that she needed. Although she had the necessary nautical almanacs and tables to use with her sextant, she was lacking a watch.

To take an accurate noon sight—meridian passage of the sun—she needed to use the sextant to watch the sun to find its highest point. This would then have to be compared to the time it occurred to determine the longitude. By only having the sextant, she could only determine the latitude she was on, which was the 19th latitude. The watch she had been wearing before the storm was nowhere to be found now.

Although unable to find her exact location, this navigation technique required a lot of mathematics. This, in turn, forced her to become more focused on surviving and not thinking about Richard. Her grief was strong, but so was her desire to live, and she knew Richard wouldn't have wanted her to give up.

Once she knew what latitude she was on, she knew she could sail to Hilo, Hawaii. The next problem was that the yacht wasn't going anywhere. The masts and the engine were beyond fixing. Although Tami had a little mechanical training, she couldn't fix the engine. She could do something about the mast and sails though.

The spinnaker pole—support for the sails—had snapped in half, leaving her with about nine feet to work with. She still had a storm jib sail available to her. She rigged up the sail to the small makeshift mast, and it was enough to get the *Hazana* moving again. While doing this, she managed to make a pump to get the water out of the cabin. Within a week, she had managed to bail out most of the water. Following on from the few small victories she had already had, she found her watch at the bottom of the bilge. Unlike the other electronics, her watch still worked. With it in hand, she could now determine longitude.

With a clear destination mapped and a ship that could move, she made her way to the North Equatorial Current. This is a westward, wind-driven current that can reach speeds of six knots or seven miles per hour. With her improvised sail, Tami managed to get the *Hazana* to travel at 2–3 knots. Sometimes she was able to complete 60 miles a day. Yet she was worried about having to leave the 19th latitude to head to Hawaii. If she didn't calculate the longitude correctly, she would miss the island and lose her only chance at survival. Despite this, her improvised sailing setup allowed her to travel 1,500 miles.

Earlier, after the capsizing, she had inflated the life raft, as she feared the *Hazana* would go down. Luckily it didn't, and she was able to access the extra supplies which were on the life raft. She didn't have many supplies, but what she did have she rationed carefully. With only a quarter tank of water, she refused to give in to her emotions and cry. This would only dehydrate her further, wasting what water she had. She also had access to peanut butter and canned food. The canned food would vary from fruit salad to sardines. She ate what she could.

Despite the small victories which had gotten her this far, the loneliness was starting to get to her. To keep herself sane, she continued to document what happened daily in her log. If anything, if she were to die, there would be a record of what happened. Although she wanted to live, her mind was fragile with the loss of her fiancé and the daily struggle to survive a little longer. Her inner voice told her to keep going, though her body was failing her. She had tried to keep her injuries clean, but they were becoming infected. When she ran out of bandages, she was forced to rip up shirts to keep the injuries as clean as possible. One of her only comforts was sleeping with one of Richard's shirts and his guitar. They allowed her to feel his presence. She also found comfort in praying, though she wasn't religious at the time.

Her mental state started to deteriorate after she noticed two ships and a small military aircraft, all of which didn't see her. Somehow, they hadn't seen her flares. She had already been adrift for 39 days when she almost contemplated suicide. She thought she had seen an island but wasn't sure if it was an island or some clouds. Then clouds gathered and she lost sight of it. It was the final straw to a mind already stretched thin. There was a shotgun in the cabin, and she was poised to use it when her inner voice told her to stop. She had come so far already, this wasn't the time to give up.

She pulled herself together and returned to the deck to look out over the horizon at the area where she had noted the island. After some time, the clouds parted, and indeed there was an island.

On November 22, 1983, the Japanese research vessel *Hokusei Maru* noticed the near disabled *Hazana* when Tami fired a flare around 4 a.m. She was just outside the Hilo Harbor. The vessel towed her inside the reef before the Coastal Guard Auxiliary towed her the rest of the way into the harbor. After 41 days adrift, Tami was finally safe. As soon as she was no longer in survival mode, she succumbed to her grief. Those who saw her were shocked at her condition. She was barely 100 pounds. She would have a long journey to recover physically, mentally, and emotionally.

Moving on to Find Love Again

The trip meant to bring Tami and Richard together had ended in disaster. Richard's body would never be found and he was declared dead at sea. Tami sent all his belongings back to his parents in Britain.

Recovery was difficult for Tami. The physical injuries took time to heal. The blow to her head had been more severe than she had first thought. She struggled to read for more than five years, suffered from short-term memory loss, and even struggled to complete sentences.

She never received any assistance for the trauma she went through and suffered from post-traumatic stress disorder (PTSD) for a long time. She regrets not getting any help sooner as she felt her healing process, mentally and physically, would have been quicker if she had. She wishes someone had made the suggestion to her, rather than trying to muscle through it. It took about eight years before she could put what happened into words. She had a lot of unresolved trauma, which surfaced as she wrote.

She wrote and self-published her survival under the title *Red Sky in Mourning: The True Story of a Woman's Courage and Survival at Sea* together with Susea McGearhart in 1998. No publisher seemed interested in her work and it took her four years to publish it. Yet in 2002, Hyperion Press picked up the book and published it worldwide. This time the title was *Red Sky in Mourning: A True Story of Love, Loss, and Survival at Sea.*

With her story becoming more well-known, it was eventually captured in the movie *Adrift* by Director Baltasar Kormákur. Although some poetic license was involved, the story stays surprisingly true to the real-life events. The book published in 2002 was reprinted in 2018 under the title *Adrift: A True Story of Love, Loss, and Survival at Sea* to tie it in with the movie.

Despite what happened, Tami never lost her love of the sea and continued to sail, though it took some time before she could. Often she will give talks to others about her incredible survival. Today, she still wears a sextant pendant with a single diamond to remind her of how she overcame all the odds stacked against her.

She even managed to find love once again when she married Edward Ashcraft in 1994. Together they had two daughters: Kelli and Brook.

Unfortunately, Kelli died in 2017 due to accidental carbon monoxide poisoning.

When asked in one of her numerous interviews how others could survive what she did, she had several things to say. Although she didn't blame the owners of *Hazana*—Peter and Christine Compton—she stated that it is preferable to rely on your own safety equipment than what is provided to you. All equipment should be waterproofed and checked before trips. Luckily, with modern advances in technology, this is something that has improved.

Lastly, don't solely learn how to navigate with satellites and other electronic devices. Because Tami knew how to navigate with a sextant, her watch, and the necessary nautical charts, she reached her intended destination. This is a must for all who plan to do an ocean crossing.

Tami is still alive as of early 2022 and lives with her husband on San Juan Island, just off the coast of Washington. Her survival was because she knew how to rely on what she had and adapted other items to give her what she needed.

Chapter 7: Lost at Sea: Steven Callahan's 76-Day Ordeal Adrift on a Life Raft

I had three cans of water left. My body and mind were shutting down; it was as if I could feel all the people who had ever been lost at sea around me. I had no more to give. –Steven Callahan

Steven Callahan was born on February 6, 1952. As a young boy, his interest in sailing started because of the Boy Scouts. By the time he was 12, he was sailing out of sight of land. By 16, he was completing solo trips. Once he reached adulthood, he had thousands of sea miles under his belt. He had had a dream since he was 12 that he wanted to cross the Atlantic Ocean. He loved boats, and when he hit his 20s, he was designing and building them. He built a 21.3-foot sloop which he fondly named *Napoleon Solo*. This was a sailboat with a single mast that contained a headsail and a mainsail at the front and back of the mast. He was going to sail the Atlantic with his boat.

His journey started in January 1981 from Newport, Rhode Island, and he was heading to Bermuda. This was a journey to start anew, as his six-year marriage had just been dissolved. This trip was meant to usher in a new beginning for him and help him heal from what had been a difficult year. He planned on doing most of the trip alone as he felt he didn't need anyone with him.

Once in Bermuda, he picked up his friend Chris Latchem, and together they traveled to Cornwell, UK. This was when Steven decided to compete in a single-hand race to Antigua. For a time, the race was going well, but poor weather hit the competitors, damaging *Napoleon Solo* and sinking several other boats. Steven limped to La Caruña, Spain for repairs and was forced to drop out of the race. He then continued to sail the Spanish and Portuguese coast, while heading to El Hierro (nickname Isla del Meridian). This island is the second smallest of the Canary Islands off the coast of Morocco, found at the most southern and western point of the island group.

His trip up to this point had been about a year. On January 29, 1982, Steven was preparing to leave the Canaries to return to America. The first week of sailing was ideal conditions, and he felt that this trip was the best thing that had happened to him. That was until a gale picked up on February 4, 1982.

The Day He Was Swept Away

Steven didn't believe the gale was going to be bad, despite the waves increasing in size and the wind blowing at about 35 knots (40 miles per hour). There are storms over the open ocean all the time; he thought it would only last for a couple of days, and then it would be smooth sailing again. He had been through far worse in the past. The night of February 4, he had engaged the autopilot and was getting himself settled into bed. He found it difficult to fall asleep as the waves bashed against his hull.

He was just about to drop off when he heard a massive crash. Within seconds the water was flooding into the cabin of the sloop. Whatever the boat had bumped into—likely a whale or large shark—had destroyed part of the hull. The water was rapidly rising, and Steven rushed to the deck to get the inflatable life raft inflated. It took some time for the raft to inflate but Steven couldn't just escape with the life raft. It barely had any provisions on it. If he wanted to survive this, he needed to go back into the sinking boat to get his ditch kit. This kit would contain most of the equipment he would need to survive this ordeal. The problem was that the boat was quickly sinking, and he didn't have much time.

Tying the life raft to the ship, he took a few deep breaths and dived back into the flooded cabin. The storm was raging above him, but below, the water was calm but dark. He had to feel his way around the cabin for his ditch bag. Going on his memory of where it was, he eventually located it. He was ready to surface. Then another disaster struck. The hatch he had used to get into the cabin had slammed shut, and the water pressure from above kept it sealed. He couldn't get out.

If he didn't open the hatch, his death was a guarantee. Then by some miracle, a wave crashed over the boat, causing the hatch to open. Taking the chance he was granted, he surfaced and managed to leave the boat behind for the life raft. Once in, the line came undone, and he started to drift away from the boat. Although the life raft had a sealable canopy, he was wet and cold from his swim. He was convinced he wouldn't make it through the night as waves continued to slam against his only lifeline.

It had been an emotional rollercoaster just getting through the initial survival. Steven had been forced to rely on instinct and training, things he had developed for years, but now it would be an endless grind to continue surviving. He had no way to communicate with the outside world, and he was at the mercy of the wind and the currents. He was about 800 miles away from the Canary Islands and steadily drifting away from them. His only hope was to float toward the shipping lanes, about 350 miles away. No one would be looking for him any time soon. He had told everyone back in America that he would be out of communication for 5–6 weeks.

Little Victories: What Kept Steven Alive

By the time the sun came up the next morning, there was no sign of *Napoleon Solo*, but at least the sea was calm. Steven took stock of the supplies he managed to gather. It wasn't much. He had about eight pints of water in cans and some food. When rationed strictly, the food could last about two and a half weeks, but the water would only last eight days. It would take him about two weeks to get to the shipping lanes if he was lucky. He needed to find ways to get water, or he would die. He compared the Atlantic to the world's largest desert. It had no end in sight, no shade, and no drinkable water.

Luckily, he found three solar stills. The problem with these stills is that they were an old design. They were used by downed pilots during World War II to create fresh water from salt water. Although they were still usable, they came with no instructions, and Steven didn't know how to use them. The first time he tried to use one, it only gave him salt water in return, completely useless.

Desperate for something to drink, when it rained, Steven tried to capture the fresh water using the canopy of the life raft. Another problem arose. The bright orange pigment used in dying the canopy had contaminated the water so badly that Steven compared the taste to drinking vomit. He refused to drink this water as he felt this would endanger his life even more.

With no fresh water being gathered, and unable to make his own with the stills, he was forced to rely on canned water. He had to carefully ration it as he had no idea when he would be rescued. Days passed, the isolation and loneliness became maddening, and he felt as if he were doomed to die. It was only exasperated by the ever-gnawing hunger and thirst. By the seventh day, Steven ripped one solar still apart, desperately trying to understand how it worked. This was when he discovered the solar still needed to be inflated in a certain way to function correctly.

With this newfound knowledge, he inflated the next still and eagerly awaited for the first fresh water to be produced. This was the first victory he had had since his boat had gone down. Unfortunately, now his food was running low. Although he had a spear gun and could hunt for fish, he had concerns about damaging the inflatable life raft he was on. He would need to be careful.

It took him a while to figure out how to use the spear gun, but by day 11, he was trying to spear the fish around his boat. It was a frustrating task, one he was failing miserably until the firing mechanism broke. The hunting implement which had given him a range of about six feet was now reduced to a common spear with a reach of only 18 inches. Frustration urged him. Unless he found a way to replenish his dwindling food resources, he was going to die.

It wasn't until the fourteenth day that he managed to spear a fish. Once dispatched, he quickly gutted it and ate what he could. Then he prepared the rest of the meat for drying within the life raft. This second victory made him feel hopeful that he would make it out of this nightmare. However, the ocean is unpredictable, and he could never afford to allow his guard to drop.

That night, likely attracted to the remains of the fish, a shark bumped along the bottom of the raft. Steven couldn't risk losing the raft. This would result in him being eaten by the shark or drowning. He defended himself by stabbing the shark every time it came within reach. He managed to connect a few times before the large fish eventually left him alone. The ocean was reminding Steven it could throw more at him.

Either late on the fourteenth day or early morning of the fifteenth, Steven noted a boat on the horizon. He immediately lit a flare in hopes he would be noticed. Sadly, the ocean is vast, and it's difficult to spot something when you are looking for it; it's near impossible to spot something if you aren't. The ship passed so close to Steven that he could smell the diesel fumes, but they never noticed him. He had managed to convince himself that he was going to be rescued, only to have those hopes dashed. Even though he had reached the shipping lanes, unless someone spotted him, he wasn't going to be rescued.

During his time on the shipping lanes, he saw several ships, but none seemed to notice him. Soon he moved through the shipping lanes, and his next possible point of rescue was the Caribbean islands, hundreds of miles away. Although averaging about 25 miles a day, life on the raft was difficult. The raft was only six feet long, making it difficult for Steven to stretch out. Exposure to the sun and water was creating seawater sores. These open sores were constantly being irritated by the salt crystals which had developed inside the raft.

After leaving the shipping lanes, he was now in the tropics. The temperatures caused his body to need more water. He was quickly becoming dehydrated, and his solar stills couldn't produce the liquid he needed to stave it off. Despite this, he continued to keep himself motivated to do what he needed to do.

On day 43, he was struck by another devastating blow. He had been hunting dorado (mahi-mahi) when the spear went through the fish. The fish then managed to break the tip off and fled, but not before puncturing the lower tubing of Steven's life raft. The sound was horrific, and Steven knew he needed to act fast, or he was in dire straits. With the way the raft had deflated, its stability was gone, and so was Steven's only way to collect water and food without falling into the ocean.

Try as he might, the patches he made for the hole wouldn't remain in place, and for the next 10 days, he was forced to use a hand pump to keep the raft afloat. He got very little rest during this time as he was forced to pump night and day. By day 51, the deflated raft was traveling noticeably slower. Steven was expending more energy than what he could take in with the rations he had. He was forced to rely on them as he couldn't do anything else.

On day 53, after pumping nearly nonstop for 10 days, Steven gave up. He was exhausted and broke down. He felt as if he had wasted his life up until this point. Yes, he was living his dream, but what did he have to show for it? A divorce, failed friendships, and a failed career. He was mentally torturing himself. Then it dawned on him. If he gave up now, he would die. There was no way around it. He was going to die. This realization shocked him back into action, and he looked through what he had. There had to be a way to keep the patch in place. Eventually, with some string and a fork, he cinched the patch in place before tying it off. The air leak was considerably better, though he still needed to pump some air into the life raft now and again. This, to him, was the biggest victory of his life to date.

Between day 54 and day 66, his food rations were nearly gone, and both his solar stills had reached the end of their working lives. The cloth within the stills had rotted away, and there was no way to fix them. The only water that Steven had left was three cans. After that, there would be nothing else, and he would die of dehydration. As the seemingly endless days ticked away, the increased salt in Steven's body was starting to mess with his mentality.

He grew increasingly concerned that he had miscalculated where the currents were taking him. If he was too far north, his life raft would take him back on the currents, further into the Atlantic, and not the Caribbean as he had hoped. There was nothing he could do but hope that he would be noticed and rescued. He willed the minutes of the day to go by.

By day 75, he had no more strength and knew his death was near. He awoke on the morning of the 76th day after he lost his boat and noticed floating plastic debris. This was the first indication that he was close to human civilization. Soon he saw the island but was distraught to see that it was surrounded by coral reefs and high cliffs. There was no safe way to land, and he was too weak to get to shore himself. This seemed to be one final loss, one that would claim his life.

Before despair could take him, he heard something. Looking around he saw a boat with three fishermen in it. He had been at sea for so long that an ecosystem had started to form around his life raft. The fish swimming around him had attracted birds that were hovering over him. The fisherman, who usually didn't venture out on this particular side of Marie Galante (southeast of Guadeloupe), noticed the birds and knew they would find fish there. One of the fishermen actually asked Steven what he was doing.

Steven was officially rescued on April 21, 1982. He had floated out to sea for so long that his leg muscles had atrophied, and he could barely stand, let alone walk, when they reached dry land. Steven had lost over a third of his body weight (40 pounds), and it took him six weeks before he could walk again. He had managed to drift 1,800 nautical miles (2,071 miles) before being found.

Making the Most of Life

Once out of the hospital, Steven would need another six weeks to rebuild the muscle he had lost during his 76 days at sea. He felt as if his life had been renewed and he was going to make the best of it. Once feeling better, he reconnected with Kathleen (Kathy) Massimini. They became sailing buddies before marrying in 1994.

Even his finances picked up as everyone wanted to know how he managed to survive out in the Atlantic Ocean. He was one of the rare survivors of the sea. He not only became a consultant to filmmakers but eventually wrote a book about his journey. In 1986, he released *Adrift: Seventy-Six Days Lost at Sea.* This book remained at the top of the New York Times' bestseller list for 36 weeks. His story was then adapted for television through the popular T.V. series *I Shouldn't Be Alive.* It aired on November 17, 2010, in Season 4, Episode 6.

This wasn't the only book he wrote. Together with James Nalepka, he wrote *Capsized: The True Story of Four Men Adrift for 119 Days*, published in 1992. This book would be revised and reprinted in 2013 under the new title *Capsized: Jim Nalepka's Epic 119 Day Survival Voyage Aboard the Rose-Noelle.*

Steven didn't let the horrors of what he went through stagnate in his brain. The life raft and solar stills he had used were less than ideal. Today, most life rafts are equipped with better solar stills. Almost as if in retaliation for the poor life raft he had been forced to use, Steven created his own. He called it the Clam, otherwise known as the Folding Rigid-Inflatable Boat (FRIB). It took him more than two decades to design. It was designed to have a hard bottom—which is rip-resistant—with a removable canopy and a sail, which allows the survivor to travel further than just the currents. The canopy was also designed to collect rainwater and offer more protection from the sun.

In 2001, Yann Martel published *Life of Pi*, an epic sea journey of a young boy lost at sea with only a tiger as a companion. This tale captivated millions of minds and eventually caught the attention of director Ang Lee. He wanted to adapt this book to be on the big screen. Many people thought it ludicrous as there was no way to correctly portray what Pi went through on his journey. Ang reached out to Steven and asked him to bring the story of Pi to life with the unique skills he had learned. Thanks to what Steven contributed, the movie wasn't only a success but felt realistic and believable. The movie *Life of Pi* was released in late 2012, allowing all of us to feel the ups and downs which Steven and Pi went through on their fateful journeys. As of early 2022, Steven is still alive and continues his adventures with his wife Kathy.

Chapter 8: Deborah Scaling Kiley's Dramatic Escape from Shark-Infested Waters

> *Every day I wake up, and it's a new day and I'm happy. I always try to find something good in the bad things that happen to me. There's never a day that you're more thankful for life than the day you almost die.* –Deborah Scaling Kiley

Deborah was born on January 21, 1958, in Texas. From a young age, she started sailing, slowly building up to finally compete in the Whitbread Round the World Race (now known as The Ocean Race) in 1981. She was the first American woman to complete the grueling 45,000 nautical miles (51,785 miles) race. This was just the start of her sailing career as she continued to sail up and down the eastern coast of America to build experience. She was a self-proclaimed adrenaline junkie.

Deborah was offered a chance to sail a 58-foot yacht named *Trashman* by Captain John Lippoth. The new owner of the boat, Morris Newberger, wanted the ship to sail from Bar Harbor, Maine, to Fort Lauderdale, Florida. The trip would take roughly six days to traverse the 1,300 miles along the Eastern Seaboard. John was in charge of gathering the crew, which resulted in a group that had never sailed together before.

The next to join the *Trashman* was Megan Mooney. Megan was no sailor and was only along for the ride with her boyfriend John. The next member of the crew to join was Brad Cavanagh. He was the youngest of the crew, and only agreed to join as Deborah was there, and she was a decent sailor. The final crew member, Mark Adams, was picked up in Annapolis, Maryland. He was an Englishman and a friend of Brad. Deborah didn't take to the sailor as he seemed somewhat condescending and chauvinistic to her. She also had concerns about John, who seemed to celebrate everything with a drink in hand.

She set her concerns aside and sailed the *Trashman* as she had been hired to do. The first day of travel went well. The four sailors and Megan got on well enough, and they were on schedule for the delivery. However, by the evening of October 24, 1982, the weather had drastically changed.

What Deborah had thought to be a squall at first was rapidly building in strength and power. The boat was pitching amongst the waves, giving all those aboard it the feeling of riding a rollercoaster. At one point, the waves were at 30 feet and the wind was gusting at 70 miles per hour. Mark had taken the helm but was drinking and hollering at the storm as if he was having the time of his life. Megan had been on the deck earlier and had taken a fall, her tether had been unable to prevent it. With no experience being on the deck while in the middle of a storm, she had been injured—severe bruising over her lower back and kidneys—and begged John to take her back to shore.

This was when Deborah realized how underprepared John was for any emergency. He had no charts to help him reach the coastline from where they currently were. A call to the Coast Guard was placed and it was suggested that the *Trashman* make its way to Wilmington, North Carolina. The ship altered its course, but as it got closer to the land, the waves grew in ferocity. Brad and Deborah took their shift on the deck, trying to control the yacht in the rough seas.

It wasn't long before the sails were damaged. They would have to rely on the engines if they wanted to make it to dry land. However, the engines soon overheated and died. Anything that operated on batteries only had limited life remaining. John was forced to radio the Coast Guard begging for a rescue, as he was concerned not only for Megan at this point but also for the boat.

Brad and Deborah finish their shift and hand over control of the yacht to Mark. They are exhausted but can barely sleep with the noise of the waves as they slam against the hull. For some reason, Mark abandons the helm by tying it off before going to sleep. He also fails to secure the storm shutters on the lee side (opposite side of which the wind is coming from) of the yacht. The *Trashman* is now at the mercy of the storm, and no one is controlling it.

The Day They Were Swept Away

While Deborah was trying to get some rest, the *Trashman* crested a wave, then free fell from it. The yacht impacted on its side, shattering the windows, and water started to pour in. The *Trashman* was going down fast. Now the entire crew was awake and scrambling for their survival.

Mark tried to undo the pressurized canister that held the life raft and all the essentials the crew needed to survive. However, he was struggling. Meanwhile, Brad noticed the difficulty Mark was having and reached for the only other option available to them. This was a rubber Zodiac dinghy. He only had enough time to grab the dinghy—leaving the onboard motor behind—before throwing it into the raging water.

He managed to hold onto the dinghy and was joined by John and Deborah.

Mark finally got the canister into the water, and the life raft inflated. Unfortunately, he lost grip of it, and the wind blew it beyond his reach. He was forced to swim to the dinghy to prevent himself from drowning. Everything they needed to survive was on the life raft, and it was gone. The dinghy was nothing more than a flotation device and had no supplies. Unless they were rescued, they had no chance of surviving this ordeal.

Soon the crew realized they were missing Megan. She was still on the ship among the rigging—a dangerous place to be when a ship sinks. She didn't have the experience of the others and had no idea how to escape her predicament. Deborah had to return to the ship to help her to the dinghy.

As Deborah watched the last of the mast disappear beneath the churning ocean, she felt loneliness she had never experienced before. The ship was gone within minutes and the five people left behind were at the mercy of the ocean and the elements. All they had to do was hang onto the dinghy until help arrived. Surely, the Coast Guard would arrive to help them soon. They just needed the strength to hang on until they did.

Surviving Sharks: Crew Battle Thirst, Hunger, and Voracious Fish

The next dawn revealed steel-gray clouds and no ships in sight. The air temperature was only slightly warmer than the water. With an air temperature of only 40°F, the crew needed to find some sort of shelter or they were all going to suffer from hypothermia. As the dinghy was upside down, it was suggested that they shelter underneath it. Using the rope and wire from the inside of the dinghy, two crew members suspended themselves with only their backs touching the water. Then, the other two lay on top of them, out of the water. Although this worked well for the crew of the *Trashman*, this option terrified Megan as she was in pain and suffered from claustrophobia. It wasn't even an ideal situation for those under the dinghy as oxygen would run out now and again. Occasionally they were forced to roll out and get fresh air before tipping the dinghy to trap more fresh air.

Although it was calm under the dinghy, the ocean continued to be rough for several hours during the day. Eventually, the crew rolled the dinghy the right way up and clambered into it to escape the water. This was when they realized Megan had been severely injured. While trapped in the rigging, it had cut her in several places along her legs, some very deeply. The crew knew that unless they got help soon, Megan's death was a certainty.

While Deborah and Mark were still in the water, Mark complained about something bumping into him. Knowing she wasn't the one bumping into him, she took a deep breath before submerging herself to see what it may have been. Sharks—and they were surrounded. The large fish had likely been attracted to them because of Megan's bleeding wounds.

The *Trashman* had gone down close to Morehead City, North Carolina. This area of the coast was known to have roughly 73 shark species that visited throughout the year. Some of the more common species include the Atlantic sharpnose shark, sand tiger sharks, bull sharks, and even the great white. The sharks identified by the *Trashman* crew were great white sharks, one of the most dangerous sharks in the world.

Normally, shark attacks on humans aren't fatal, but the violence of the attack is terrifying. People are often bitten by sharks because they enter murky water and thrash around on the surface. The shark views this activity as a potential prey item that is injured. As they are opportunistic feeders, they will attack in hopes of an easy meal. At this point, with blood in the water, the great whites had come for a free meal. Mark and Deborah were in danger.

The two of them hurriedly climbed into the dinghy. As soon as they were out of the water, the fins of multiple sharks started appearing. Deborah had never seen so many sharks gathered in one place before. It was only by some miracle that they hadn't been attacked during the night before. However, now that they were all aboard the dinghy, the sharks circling them kept trying to nudge the boat. It was almost as if they were hoping to tip the boat over and get to the meal eluding them.

The crew needed to stabilize the dinghy. After some searching, the crew managed to find a piece of plywood. They tied it to a rope to the outside of the dinghy and threw it into the sea, hoping they had created a sea anchor. Instead of stabilizing the dinghy, a shark took hold of the wood and started to swim away with it, much to the horror of the humans. Eventually, the shark lost interest in the wood and let it go. The shark had to have been at least 700 pounds to pull the dinghy and five adults at the speed that it had.

The storm which still had them in its grip eventually died down at nightfall. There had been some rain, which gave them some relief from their thirst, but it didn't last long. Everyone was hungry, cold, and wet. They were also always surrounded by sharks, although they seemed to come and go as they pleased.

By the time the third day dawned, the water in the dinghy was fetid. It was a combination of pus from Megan's wounds, urine, and blood, and seaweed. Because of this, everyone was breaking out in *Staphylococcus* infections, and their skin became sensitive even to touch. Megan's wounds were so severe that she was now suffering the effects of septicemia, and there was nothing anyone could do for her.

During this time, Brad and Deborah formed a bond to help each other survive. It was clear Megan was beyond help, and John and Mark were starting to get delirious. Later in the day, the two of them would turn to drink seawater to sate their thirst. Although this may have quenched their thirst initially, the cursed liquid was already going to work. Seawater has high concentrations of salt. By drinking it, the body tries to expel the excess salt by drawing more water from what reserves the body has left. This causes the kidneys to become damaged. Unable to remove the excess salt from the body, the hapless victim becomes delusional. Death arrives soon after, either from dehydration or the victim playing out a delusion that kills them.

Because of the drinking Mark and John had done throughout the trip, they were likely more dehydrated than the rest of the crew. This caused them to turn to seawater to sate themselves. By the fourth day, Mark and John had descended into madness, mere hours after drinking seawater. John was the first to act out a delusion. Not only was he convinced he saw the land, but tried to convince the others he was going to get his car. With no energy to stop him, the crew could only look on as John climbed out of the dinghy. Try as they might, no one could convince him to return, and soon they floated away until they could no longer see him. They could still hear him when he was attacked and devoured by the sharks, which had never left.

Next was Mark. He had been babbling incoherently for a while before stating he wanted to go get beer and cigarettes. He also stated he wanted to stretch his legs a little. He too went over the edge, before being dragged under the dinghy by the sharks which had already been fed once that day. No one had the strength to look away as the ocean boiled with the frenzied sharks and blood. Deborah considered this the worst moment of her life.

That night, Megan started acting strangely. She was waving her hands in front of her face and babbling in tongues. Likely, at this point, the blood poisoning was in its final stage, and she was dying. Brad and Deborah had no choice but to watch the poor woman die before them. Eventually, they managed to fall into a fitful sleep.

On the morning of the fifth day (October 28), they found Megan had died during the night. Mind clouded by hunger, Brad suggested they use Megan as nourishment. The act of cannibalizing a deceased person isn't shocking, as many survivors have had to turn to this resource in dire situations. However, Deborah put a stop to that idea. Megan had likely died of blood poisoning, and eating her would only seal their fate. They also couldn't let her corpse remain in the tiny dinghy as it would start to decay and make them more ill. Instead, they stripped her body of clothing and jewelry with the intent of returning it to her family if they survived.

Despite Megan being dead, the pair of survivors couldn't bring themselves to just throw her in the water without holding some sort of funeral. Together they sang Psalm 23 and said the Lord's prayer before delivering Megan to the depths of the ocean. Unable to bear the sight of the sharks getting to her, Brad and Deborah forced themselves to go to sleep.

They woke a few hours later and realized they were going to get sicker if they didn't clean the fetid water from the bottom of the raft. Deborah convinced Brad to do this, as she barely had any energy left. Brad was no better, but he tried, despite his terror of the waiting sharks. Then disaster struck. As he tipped the boat, he fell over backward into the water. Terror gripped the both of them, but Deborah had nothing more to give and couldn't help Brad back into the boat. She sat against the side and cried to herself. She had doomed Brad.

Brad wasn't ready to give up, and with the last bit of strength he had, he managed to pull himself back into the dinghy. For a time, the two of them sat at either side of the dinghy, ignoring the world around them. Then Brad noticed a ship, and he tried to convince Deborah of what he was seeing, but she didn't believe him. She felt that he was losing his mind because of the dehydration. However, he wasn't hallucinating, he really did see a ship, and it was heading right toward them.

Somehow, the Motor Vessel (MV) *Olenegorsk*, a Russian freighter, had spotted the dinghy and came right up to it. The passengers of the dinghy waved at the men on the deck and they waved back before dropping a lifesaver in the water for them. With a real rescue in sight, Deborah leaped from the dinghy and swam for her life. Brad, still terrified of the sharks, weighed his options before following.

Despite 1982 being the height of the Cold War, these Russian sailors saved the two Americans before contacting the Coast Guard to collect them. The two of them had floated over 140 miles from where Trashman had gone down. They were steadily heading toward the open ocean and certain death.

It was with horror that Deborah and Brad learned the Coast Guard had called off the search for the *Trashman* soon after it had gone down. Someone—who has never been identified—placed a call stating the *Trashman* had made it safely to the harbor. If this call hadn't been made there could have been a chance to save everyone. They were taken to Morehead City for treatment.

Motivating Others Through Words and Images

Deborah and Brad both made a full recovery and went on to live their lives separately. Brad went on to become a professional yachtsman and Deborah a motivational speaker. Although a defining moment in her life, Deborah didn't want the tragedy to define who and what she was. Eventually, she managed to work through the trauma to publish *Albatross: The True Story of a Woman's Survival at Sea* in 1994. This would later be republished under *Untamed Seas: One Woman's True Story of Shipwreck and Survival* and was written by Deborah and Meg Noonan in 2001. It underwent one more rerelease in 2006 by Deborah. This time, not only did the name change, but it also included photos taken of Deborah and Brad after they were rescued. *The Sinking: One Woman's True Story of Survival at Sea* was published in 2006. She then wrote another book in 2006 titled *No Victims Only Survivors: Ten Lessons for Survival*, explaining what she had learned from the sinking of the *Trashman*.

With the first publishing of her first book, she originally didn't want to go through with it. She felt that she was earning blood money as three people had perished. Someone eventually convinced her to do it, but then to donate some of the proceedings to a charity. This made the ordeal better for her to bear.

It wasn't only pages of books that told her story. On October 28, 2005, the television show *I Shouldn't Be Alive* aired her survival story as the pilot to their show. Together she and Brad told their tale of survival. Then in 2019, Shark Week released a T.V. movie based on the events of the sinking of the *Trashman. Capsized: Blood in the Water* was an instant hit, and people wanted to know more about the duo who had survived.

Although she had survived a terrible tragedy, more seemed to follow her after her survival. She was married and divorced twice, first to John Coleman Kiley III and then in 2005 to Greg Blackmon. The latter marriage dissolved in 2008. She had two children with John: a daughter named Marka Kiley and a son named John Coleman Kiley IV. Sadly, her son died on August 23, 2009, due to an accidental drowning. He was only 23, a little younger than his mother when she was shipwrecked. Sadly, Deborah passed away in San Miguel de Allende, Mexico, on August 13, 2012. She was only 54. No cause for her death was ever released to the public.

Before her death, she was asked how she had managed to survive. She put it down to three things: her hope and faith in God, visualizing herself surviving the ordeal, and having the unshakable will to survive. She felt the problems with the boat stemmed from poor choices like drinking saltwater and lack of proper preparation. The boat wasn't equipped with the correct charts and there was a passenger who wasn't trained to deal with emergency situations. Although Deborah is no longer with us, her words and actions still are, helping those who have also managed to survive terrible odds.

Chapter 9: José Salvador Alvarenga: The Epic Story of the Longest Solo Survival at Sea

> *I suffered hunger, thirst and an extreme loneliness, and didn't take my life. You only get one chance to live—so appreciate it.* –José Salvador Alvarenga

Before José became known for the longest adrift at sea, he wasn't known for being the most upstanding person. Originally, he was from El Salvador but had left the country in 2002. He left behind his mother, María Julia Alvarenga, his father, José Ricardo Orellana, a girlfriend, and his young daughter, Fatima. He settled in Mexico to become a fisherman. He rarely, if ever, contacted his family as he continued to eke out an existence as a fisherman.

During this time, he had a fairly poor reputation and was known for his overindulgence in everything from women to parties and food. This earned him the nickname Chancha, or pig. Despite the terrible nickname, he was well-known and beloved by those who he worked with. He was an experienced captain (having been on the water for many years), an experienced fisherman, and even had a few stories about hairy situations he had gotten himself into on open water.

In 2012, he was working for Bellarmino Rodriguez Solis (also known as Willy) for several years and was comfortable with his job. José planned a fishing trip for November 17, 2012, and wanted his friend Ray Perez to join him. He had worked with this man in the past and trusted him. Sadly, that day Ray was unable to join him. José, instead of giving up on the fishing trip, invited Ezequiel Córdoba (nicknamed Piñata). The two were practically strangers, and although inexperienced, Ezequiel jumped at the chance to work for some extra money.

The two set off in a boat about 25 feet in length and half that in width. Going by the name of *Camaroneros de la Costa*, this boat had no raised structure on it, no running lights, and no glass. This was the kind of boat you would spend a day out at sea and return as soon as possible. However, on November 17, 2012, a storm surprised Ezequiel and José, resulting in them being washed out more than 50 miles to sea.

The Day He Was Swept Away

At least the men were prepared. They had had a good day catching sharks, mahi-mahi, and a host of other fish, all of which were in a large fiberglass crate. They also had a host of equipment which included items such as a mobile phone in a dry plastic bag, knives, buckets, a two-way radio, and a GPS tracking device.

Despite struggling through the night in the storm, the two of them had enough fuel to get back to land. José was able to prevent them from capsizing as he battled throughout the night to keep them on course back to Costa Azul, Mexico, where they had left a day ago.

Ezequiel was in charge of bailing out any water which landed in the boat.

By 9 a.m. on November 18, 2012, there was finally a break in the weather, and José noted a few mountain tops in the distance. They would be home soon if the weather remained calm. Then disaster struck and the waves damaged the engine, leaving them stranded dead in the water. José radioed Willy and explained their situation. However, as the GPS had also been damaged in the storm, José had no way to give their current coordinates to Willy, despite him wanting to. Willy suggested they anchor as the worst of the storm was over.

Then José admitted he had knowingly left without an anchor, as he believed they weren't going too far out in the water. Communication with Willy didn't last much longer, as the radio died around 10 a.m.

Waves continued to rock the boat, and José was worried the filled fiberglass container would tip them over. He ordered it to be emptied. Together with Ezequiel, they tipped over 1,000 pounds of fish back into the water. Although this helped stabilize the boat, it would now attract sharks. They would need to be careful and not be washed overboard. Thinking quickly, the two men bound together the buoys of the boat to create a sea anchor to aid their stability and slow down the pull of the currents.

They continued to battle as the weather worsened once more. Throughout the night they would take turns sheltering in the fiberglass container and bailing water out every 15 minutes. While they did this, they were slowly pushed away from land and possible rescue. The following morning, the storm was gone, and they could no longer see any land. Most of the equipment was useless or had been washed out to sea.

Desperation, Willpower, and Imagination: Skills for Survival

Days passed, and no rescue came. The men became desperate for food and water. They had lost all the lines, hooks, and bait, which could have saved them from this situation. However, José wasn't willing to give up. He developed a technique where he caught fish with his hands. It took some time, but eventually, he caught food. These fish would either be eaten raw or dried. Sometimes they were lucky and caught turtles.

Both knew they shouldn't drink seawater but did turn to drinking urine. As the kidneys are used to remove excess salt from the body, they are still consuming more salt than they should have. Now and again it would rain, and they collected enough water to drink their fill—using the crate and buckets—before rationing what remained. The pair quickly learned to scavenge what the ocean gave them, even if it was garbage bags filled with trash. Occasionally they would see ships at a distance, but none would ever come close enough for the men to be spotted. They also had no way to communicate to signal the ships as their boat was almost perfectly camouflaged into the ocean.

After about two months, José had adapted to their new lifestyle. Sadly, Ezequiel wasn't doing as well. He was mentally and physically drained. Then, following an incident where he ate some raw seabird—which José referred to as a duck—he became ill and refused to eat any more. A combination of depression, a refusal to eat, and barely drinking was causing Ezequiel's body to shut down. Eventually, too weak to do anything, Ezequiel died after roughly four months out to sea.

José was distraught. He had invited the young man to join him on this trip, and now he was dead. He propped Ezequiel up to prevent him from falling into the water and continued to talk to him as if nothing had ever happened. In time, Ezequiel's body started to mummify in the heat, and by the sixth day, José knew it was time to set his body free.

Once the ocean claimed Ezequiel's body, José fell into his own depression. For a time, he considered suicide but was scared of going through with it, as he feared he would never go to heaven if he did. Instead, he kept himself sane by using his imagination. He would go for long walks on the boat while daydreaming he was somewhere else. He also imagined vastly different meals from what he was forced to consume when he could catch them. This likely caused him to think he was hallucinating when he started noticing shorebirds.

He had been out to sea, by his calculation, for 14 months before he noticed the birds. Looking around, he was rewarded with an island in his sights. It was a small atoll, but an island was better than remaining on the open sea. Luck was on his side as the currents caused his boat to drift closer to the atoll. Once he was within 10 yards of the island, he leaped from the boat and paddled toward it.

José had managed to drift all the way to one of the 1,156 islands in the Republic of the Marshall Islands. This particular atoll was Tile Islet, forming part of the Ebon Atoll. It was also the home of married couple Emi Libokmeto and Russel Laikidrik. Once José managed to drag himself from the shore, he crawled through the undergrowth until he was discovered.

Despite the language barrier, the couple cared for him before Russel went to the main town located on the island of Ebon. He was hoping the mayor could offer some help for what they thought was a half-mad man. José was officially rescued on January 30, 2014, after 438 days adrift on the Pacific ocean.

Life of Controversy

From the state of his body, it was clear to everyone that José had been at sea for a long time. Yet, for how long, no one was sure. Since his rescue, José has been shrouded in controversy. Some stated he looked too healthy, had no sunburn, and didn't seem to be suffering from any of the known shipwreck diseases when he was rescued.

He didn't want to talk to the press about what happened to him during the 11 days he was in the hospital. During this time, only his doctors could speculate what he had been through by looking at his body. José was suffering from anemia and had a damaged liver. This damage was likely from following a raw meat diet for as long as he had, ingesting many parasites.

Once in hospital and receiving fluids, his legs rapidly started to swell. This was a sign that his strained body was trying to absorb as much water as possible, due to him not having enough fresh water for an extended period. His kidneys were fine, which proved he had had some form of fresh water during his travels.

Due to him not being severely sunburned, his kidneys were undamaged, and he didn't suffer from scurvy, many people believed he had lied about the duration of his drift. Even those who had been part of the search and rescue, which concluded after two weeks, couldn't believe José had survived. Some people even went as far as to say he had stopped off on multiple islands during his misadventure. Whether he stopped off on other islands only he will know, but as for the duration of being adrift, there is no denying the truth. Multiple people witnessed him and Ezequiel leave Mexico on November 17, 2012, and heard the SOS call placed with Willy. The ship Ezequiel and José left on was the ship that had left Mexico and was found in Ebon Atoll.

As for not suffering scurvy and fatal dehydration, José's diet of sea turtles explains that. These animals have a lower level of salt in their blood and contain vitamin C. In combination, this would have staved off scurvy and dehydration for a time. While being out to sea for 14 months, José was likely able to capture enough water from rain to prevent himself from dying of dehydration. The fiberglass crate also protected him from the sun, which likely contributed to him not having severe sunburn when he was found.

The state of his body couldn't be faked, and many experts confirmed that what he had gone through was plausible. As for how he got to the atoll—which seemed to be a straight line for 6,700 miles from where he left—it was determined that the sea currents carried him. Had he missed the islet, he could have drifted as far as Papua New Guinea or even Australia, if he managed to survive that long.

The last controversy was that José was accused of eating Ezequiel and not allowing the ocean to claim him. Cannibalism for survival isn't something new. The most famous story is that of Uruguayan Air Force Flight 571, where cannibalism saved some of the passengers while they waited for rescue for two months. José denied the allegation and insisted he had treated Ezequiel's body with respect.

While recovering, José spoke at length with a journalist named Jonathan Franklin about what had happened to him. Jonathan went on to publish *438 Days: An Extraordinary True Story of Survival at Sea* in late 2015. He used the information provided by José, those who nursed him back to health, and many other professionals to paint a picture of what José had gone through. The dates within the book were rough estimates according to José's memory when he told his tale to the journalist.

Whether the world believes his tale or not, José is a man who left Mexico on November 17, 2012, and was only rediscovered on January 30, 2014. He had faced horrors most people can't even dream about and managed to make it back to civilization. His time out to sea scared him so badly that he feared being near water for a long time. He returned to El Salvador and was reunited with his parents and young daughter. He is even striving to be a better father to help build a relationship with his daughter.

Chapter 10: Michael Izquierdo and Crew: Untold Story of My Father's Survival Journey

> *It was in this moment that I had come to a realization that nothing is worth this, money is not everything.*
> –Michael Izquierdo

Michael Izquierdo has been a mariner all his life. He was born in 1969 and was the eldest of four boys. Growing up, he had set his sights on only one profession; his greatest ambition was to be the captain of his own ship. He had always been fascinated by ships and longed to navigate one. By the time he was 20, he started to lay the foundations for his dream.

As he worked his way through the different ranks of officers, he took the time to get married at 25. He ended up having three children with his wife Brigette. Despite this happiness, life as a marine officer on a vessel was hard work, physically and emotionally. He would spend roughly nine months out at sea, only to get two months with his young family before returning to the sea for another nine months. When he was on the ship, his life revolved around it 24-7, and he had little time to think of anything else.

As it was the early 90s, his only form of communication with his family was to write letters, as the internet wasn't as prevalent as it is today. For years he lived this way as he studied and worked hard to complete the various exams which would help him achieve his dreams. Eventually, his hard work paid off, and he made it to Chief Officer (Chief Mate). This was only one step away from where he dreamed of being. The captaincy position was just outside of his grasp. He was almost there.

In 1997, an opportunity arose that would change his life forever. He had finally completed his qualifications to be a captain. The problem was that no one had offered him a captaincy contract. Instead, he had once more been offered a nine-month contract through the last agency he had worked for as the Chief Mate. Knowing it was just a matter of time before he got the role of captain, he considered taking this contract, when another offer stopped him.

This new offer gave him the position he longed for. He was going to be Master Mariner (Captain) for a small cargo company that was sailing the Asia-Pacific region. This was the opportunity he was waiting for! The ship was older than he would have liked, but it was the first time he was in command. Also, the pay hike wasn't something he could ignore. Being young, ambitious, and filled with excitement, Michael accepted the offer with confidence. He had earned the right to be a captain, why not take this opportunity? This was going to provide a brighter future for him and his family.

At first, the job was amazing. After four months on the Asia-Pacific route, he had delivered goods to Hong Kong, China, and South Korea. The longer he worked as captain, the more he knew this had been the right call. That was until one stormy night changed his life forever in September of 1997.

The Day They Were Swept Away

"The ship is listing three degrees on the port side." He had told his crew as they sailed.

He had first noticed this odd tilting to the left around 9 p.m. on September 19, but hadn't been overly worried about it. To ensure it was nothing, he asked his Chief Mate to see what could possibly be causing it. As the ship was explored for the possible cause, nothing stood out at first. The ballasts were first checked for signs of water, but there were none. Other crucial areas of the ship were also checked, but the crew came up with nothing. Eventually, the cargo hold was checked, and it was here that Michael's worst fears were realized.

Water was flooding into the cargo hold. At this rate, the kaolin clay—an ingredient used to make porcelain—was mixing with the rising seawater, ruining it. Desperate to hold the ocean back, a submersible was used to help pump the water out of the hold. However, this wasn't enough, the listing became worse, and soon they were tilted by eight degrees. Now Michael was concerned, but he couldn't show it to his crew. He needed to remain optimistic and continued to have the water pumped from the hold.

They were only two days away from Manila, Philippines, their next stop. The MV *PROMEX BAYU* just had to keep going a little longer before they could be safe. However, while they were still in international waters within the South China Sea, they were surprised by a storm. This was one of the worst possible things that could have happened. Michael quietly wished his ship would make it to the harbor, the same wish he had made in the past when his car's gas warning light went on when he was only a few miles away from a gas station.

The crew would spend the rest of the evening until the morning of September 20 pumping the water from the ship. Suddenly, the engines died. The ship had finally tilted too much, taken on too much water, and was no longer safe for those onboard. They were stranded, unable to move, while a storm raged around them. Soon, the ship was listing by 45 degrees, and Michael was forced to send out a distress signal to any ships in the area. The first people to pick up the signal was the Manila Station, but for some reason, never sent a search and rescue team to Michael's location. It was likely they didn't have the resources at the time to help. Despite that, they alerted the crew's shipping agency in hopes that they could communicate with the ships in the area to help their crew aboard the *PROMEX BAYU.*

Failing Ship: How Michael Rallied His Crew

Dread gnawed at Michael's belly. Waves as tall as 26 feet and gale force winds pounded at his ship. It could be hours before they were rescued and he needed to think about the lives of his crew. He forced himself to remain calm, though he noticed his crew was starting to show signs of nervousness. He called them together just after 10 a.m. that morning and explained that they would need to evacuate the ship. They all needed to prepare the lifeboats in case help arrived too late.

Their ship was equipped with two life rafts and two lifeboats, roughly the size of a 4x4 pickup truck. Preparations to get the first life raft into the water quickly got underway. These rafts needed to be released into the water first before the crew would descend a ladder to board them. However, when the first life raft was launched, the crew failed to realize the rope holding the raft had snapped. With nothing to hold it in place, the storm snatched it away before they could use it.

Disbelief and more panic gripped them, but they immediately moved to the second raft on the starboard side. Time was running out as the ship was listing more and more, sinking faster. Their second attempt didn't fare better than the first. As they got the raft into the water, the rope snapped again, and the strong winds took hold of the life raft, driving it farther from the ship.

Now the situation was turning dire. Two life rafts had been lost to the sea, and there were only two lifeboats left for 21 men. Despite the chaos around him, Michael kept a cool head. He pressed his crew to launch the first of the lifeboats. Paranoid that the sea would steal their lifeboat once more, the crew ensured the ropes kept it in place while working. They needed to be careful, anything could go wrong, and then they would only have one more chance to get off the ship.

The ship was foundering fast and Michael made a split decision to have his entire crew board the one lifeboat. Once it was launched, he ordered his men in one by one until only he remained on the ship. He feared for his men, as the waves could easily topple their lifeboat or slam it into the sinking ship. Luckily, none of that happened. By the time it hit midday, he and his entire crew were squashed into the lifeboat. The weather was so foul they never saw what happened to their ship in the end.

It had been hours since they had sent the distress signal, and they were in dire straits now. The lifeboat was open to the elements, and they had no form of shelter. What it did have was a radio and an emergency position indicating radio beacon (EPIRB). With this, any ships close by would be alerted to their location. They immediately turned it on and started hoping for rescue. Their ordeal was far from over.

Waves 16–26 feet slammed their small boat from every direction, and having 21 people in it was not ideal. Michael ordered his crew to hold on to anything they could to prevent them from moving in fear of overbalancing and capsizing the small boat, dooming them all. They just needed to hold on.

An hour later after they were crammed into the lifeboat, a plane flew over them. They didn't think much of it until it hovered over them, and their radio crackled to life. The plane above them shared that a rescue team had been dispatched from Hong Kong, and was on its way to their location. The voice then asked them to shoot a flare, so it would be easier for the rescuers to see where they were. They weren't out of the woods yet, but help was at least on the way.

Despite this, Michael couldn't help but consider what they had been through. As captain, he was responsible for the men who huddled around him. With the waves still pounding at them, anything could happen, and the young captain was starting to fear for his life. With no ship to control, at the mercy of the elements, he felt the gravity of their situation. He thought of his young family. Would he ever get to see them again? Was this the price of his ambition? No amount of money was worth putting his life in this kind of danger.

It was nearing 3 p.m. when the stranded men spotted the approaching cargo vessel. This vessel had received the distress call from the Manila Station and had been guided to the crew by the EPIRB locator beacon. The entire crew of the *PROMEX BAYU* had been rescued with little to no lasting injuries.

The Korean vessel helped the bedraggled crew aboard before giving them dry clothes and something warm to eat and drink. A few hours later, the 21 men were transferred to a Philippine Navy Vessel. This ship brought them back to shore where they received medical attention before being released to be reunited with their families.

This left only one question unanswered. What happened to the *PROMEX BAYU*? When you try to do research online, its name rarely pops up in any searches. Due to the time the ship went down, there were no camera cell phones, and it wasn't documented by the crew. All in all, the sinking of the ship didn't even make any news headlines, which is quite common for many vessels, especially when no official investigation was launched. In the end, the sinking of *PROMEX BAYU* was written off as *force majeure*. Its sinking was considered an act of God, and no one could be held responsible.

Meanwhile, the truth behind the ship's demise was well known to Michael. He had been privy to the ship's history when the previous captain had handed him the keys. This ship was so old that it had technically already been decommissioned when it sailed under the flag of the current owner's country. Michael had learned that the licenses for the ship had been revoked by the governing maritime authority in the past. However, the company had circumvented this by not only forging the necessary paperwork to make the ship seaworthy again, but also by registering the ship under another flag from a different country.

Despite being stopped in multiple ports for failing quality checks, the company had just promised to amend the problems within a few months. They never did. The reason the PROMEX BAYU sank was because of a crack that could no longer be patched. However, as he was still young and had a family to worry about, Michael went along with the company's statement of the cause being force majeure. Soon after, he refused to work with the company and cut ties.

Rewarding Sailing Career

Although this experience changed Michael, it didn't deter him from pursuing his dreams in this profession. He still had a family to support. He continued to work hard to become more skillful in his craft. Except this time, he insisted on the safety of all of the boats he served on until the end of his mariner career.

Since the years passed after the incident, health, safety, and quality control have greatly improved in the maritime industry. Because of this, Michael would continue to be an exemplary Master Mariner of all the cargo vessels he stepped foot on for the next 25 years. Sadly, he eventually had to retire and did so in February 2022.

With more time on his hands now, he looks forward to enjoying his life together with his wife, Brigette. Together they travel to New Zealand to visit my small family. I am not only grateful that he got to live out his dream but also for his actions on the day his ship sank. Thanks to his quick thinking and keeping a cool head, my children get to experience the sea through his storytelling.

Bonus Chapter: Improving Your Odds of Surviving a Maritime Disaster

As with planes, you can give yourself an edge in surviving a shipwreck. You even have a bit more freedom when it comes to what you may and may not pack. When on a cruise, there are a variety of items you can pack, which you wouldn't necessarily have access to when on a plane. This makes packing your own ditch bag possible. However, before you decide to add a life raft to the list of items you want to pack, do the necessary research about the cruise you're on. Some items you may want to take could be banned, and there is no reason to have your own life raft when one will be provided in an emergency.

Having usable items such as a flint and steel, a torch, a first aid kit, and so on in a sealable, watertight container is a great idea. However, for surviving a shipwreck, it's more about what skills you have that will allow you to survive the longest. Knowing how to swim is a necessary skill, as is knowing how to build a shelter, start a fire, and find fresh water. Some of these skills can be used on the life craft, while others will help you if you reach a desert island.

Surviving Cruise Ships

The best way to survive a shipwreck when you're on a cruise ship is by paying attention. According to the SOLAS regulatory rules, all cruise ships should have a muster drill within 24 hours of the ship leaving port. This drill is to educate people on what to do in an emergency. This includes identifying muster areas (emergency gathering areas) and teaching you how to use a life vest. Both of these are crucial.

Some cabins will have life vests, but you don't have to wear them all the time unless otherwise stated by a crew member. When at a muster area during an emergency, life vests will be handed out.

Muster drills are compulsory for everyone, even if you have sailed before. Not only is this a safety briefing, but extra information about safety procedures will be given to passengers. Don't be afraid to ask the crew any questions if you're concerned about where vital equipment—such as the first aid box—is on the ship. Familiarize yourself with the layout and know what you have to do when the emergency alarm goes off. Most importantly, don't panic. This causes more deaths than anything else.

As a passenger on a cruise ship, there is little you can do when it comes to getting the lifeboats ready for evacuation. The best you can do is prepare yourself and listen to all instructions. The crew have been trained to survive and will teach you what you need to know.

Surviving Personal Ships

When you own a personal ship, it's vital to ensure it's always in excellent condition before going out to sea. This isn't just about checking the hull and ensuring there aren't any leaks. You need to have all the equipment—like electronics, a radio, and a ditch bag—tested and be in usable condition before leaving the land. Once any boat is out over open water, there is no way to return easily to get something. Ensure that you have a working EPIRB on your boat, especially when traveling vast distances.

Before setting out, make sure you know what the weather is going to do. When too bad, skip fishing or make sure your vessel is ready to weather a terrible storm. Not all vessels are created equally, so know what yours is capable of before putting it to the test in rough weather.

Know the area you'll be sailing in. This includes having the correct charts and maps available to you. Plot your destination and know what potential dangers may present themselves while you're on the water.

Inform everyone who will be traveling with you that they have to wear life jackets—no ifs, ands, or buts. When you are the captain of the vessel, your job is to keep the ship floating and your passengers safe. Ensure everyone is correctly fitted to their life jacket and they know all the emergency procedures of your vessel.

The most common reason ships get damaged is that they impact something. Know how to patch your ship if it springs a leak. Get people to help you if you cannot leave the helm. Use clothes, towels, anything absorbent to keep the water flow to a minimum until you can view the damage yourself. Depending on the boat type, you may be able to tilt it to bring the damaged area out of the water.

An impact can also result in injuries among passengers. Ensure you have a fully stocked first aid kit and know how to use it. After an impact has occurred, radio the Coast Guard or similar authorities in your country. The important information you will need to give them includes your location (coordinates), the name of the vessel, and what kind of emergency you're experiencing. You can even describe what the boat looks like so it's easier to find. At night, your best bet is to use torches or flares to get someone's attention. If the vessel can still move, get it to land as quickly and safely as possible. Ensure your boat is equipped with a bilge pump to help pump excess water out of the boat while you try to get to land. When a leak is minor, address it as quickly as possible, as every second counts as water floods the boat.

Lastly, know when there is no hope for your boat and you need to evacuate. Tell the Coast Guard you will be evacuating and prepare those with you to get in the life craft you have available to you. Take necessities only! Ensure you have an emergency radio, enough freshwater and food to last everyone, a first aid kit, and a working cell phone. This may be easier if you're on your own but will be more difficult if you have panicked passengers or children. Keep them calm, give strict orders, and get other calm people to work with you to achieve the evacuation process.

Overall Survival

Your best chance of survival is to keep the boat afloat with you in it. You'll have a higher chance of survival if you do. You will have a lower survival chance on a life craft and even less if you're in the water. When it comes time for you to evacuate the ship, remain as calm as possible. When on a cruise ship, follow all the instructions given to you by the crew. When on your own boat, get the life craft ready to get people off the boat. Ensure these crafts don't blow away, as these may be the only lifeline you and your passengers will have before rescue arrives.

On Sea

As with when on land, you need three essentials to survive: shelter, water, and food. Finding shelter out in the ocean is near impossible unless you have the correct life craft to rely on. Many modern life crafts come standard with a canopy that will protect you from the sun and the rain. Exposure to these two elements will kill within hours if they are extreme enough. Wet clothes should be exchanged for dry ones when possible, as this can exacerbate hypothermia.

When a canopy isn't available, create one with what materials you have. Use blankets, extra clothing, and even a sheet of plastic tied to the four corners of the life craft to keep the heat and cold off of you. Recognize symptoms of hypothermia and sunstroke and treat accordingly. Hypothermia requires a person to be dry and given a source of heat. When blankets aren't available, share body heat. When suffering from sunstroke, the person needs to be cooled. Don't allow them to swim! The ocean is filled with danger, and there is a chance debris in the water can cause further injury. Take restricting clothing off and dunk shirts into the water before applying it to the neck, groin, and armpits. No matter how hot it gets, avoid swimming. This will only drain your energy reserves. If you don't have food and water, you will not be able to replenish your energy. It may also become difficult for you to get back into the boat, or you'll topple it, sending other survivors into the water. Keep activities low to avoid losing too much water.

When the Coast Guard is called for assistance, help is sent out, or ships are notified about the sinking. Try to remain as close to where the ship went down as possible. This is the last location the inbound rescue has of you. To avoid floating too far away, you can create sea anchors of buoys and other floating debris to slow the rate at which the current pulls you away. This isn't always possible. This is why having an EPIRB is vital to you being found.

The only way to have fresh water on your life craft is by bringing it with you, having floating solar stills, or catching rainwater. You cannot last long without water. Dehydration is the second on the list of things that can kill you while stranded out in the ocean. Ration water if you have a limited volume. When not actively perspiring, the feeling of thirst can be eased by sucking on a button. It generates saliva, which will quench some of the thirst. However, this will not work for extended periods, as the human body needs 50–68 fluid ounces of water a day to maintain its condition. You must avoid drinking seawater at all costs—this includes wetting your mouth with it. You will also need to avoid drinking urine as it already contains salts your kidneys have removed. Generally, drinking blood is also a bad idea, but if you can grab a turtle it'll be a lifeline in terms of food and water. Their blood is drinkable, and the meat will stave off scurvy.

Getting food will be dependent on what you grabbed, how long you're at sea, and which sea you're in. The longer the life craft is out to sea, the larger the ecosystem that develops around it. This will give you access to fish, turtles, and even birds. The problem is trying to catch these creatures without getting injured or toppling from the life craft. The easiest way to catch fish is with a net. You can achieve this by making one with any rope you have with you or using a shirt with one end tied closed. Alternatively, if you have some debris that you can fashion into a spear, this would also work but may be more difficult to fish with. To hunt with a spear, you'll need to practice. The water refraction will cause the fish you're after to appear in a different location from where you think they are. Before hunting, add the spear tip into the water to see how it bends. This will give you a rough idea of how water refraction works. When hunting fish, avoid those with beaks, spines, or those that can puff up. Generally, these will have poison that will kill you.

If there are any lessons you can take from the survivors in this book, it's that you have to be resourceful. Nails and wires can be bent to be hooks. Ropes can be strengthened to make fishing lines strong enough to even catch sharks. As a survivor in the middle of the world's largest desert, you must learn to become a creative scavenger if you want to live. Even a floating green coconut has value. The liquid within can be drunk, while the flesh can be mashed up to create an effective sunscreen.

Although not considered by many as essential, morale is vital to everyone's survival. The lower the morale, the higher the depression, and this causes people to give up. It can also cause people to act irrationally, placing others in danger. Do what you must to keep your morale up.

On Land

Distances at sea are deceptive, and it's easy to see clouds on the horizon and believe them to be an island. Don't leave your life craft to swim to an island you can't confirm up close. Some signs that you're close to an island are the change in water color and seeing more shorebirds.

Although not always true, an island is generally safer than being in the open ocean—especially if it can provide you with fresh water and food. The other advantage is that you're no longer moving, making you easier to find for rescuers. The disadvantage is that you aren't moving, making you more difficult to spot. This is why you should still think about the three essentials of survival and a way to attract help.

Beacons can be made with messages in the sand (classic SOS), fires burning black smoke, or flares when ships or planes are in view. Any fires you make will be a blessing. Fire can be used for light, cooking, protection, and sterilizing. Alternatively, if you have a cellphone (even if it no longer works), a mirror, or a piece of metal that's reflective, you can signal passing ships and planes. By moving the reflective surface back and forth, you catch the sun and create an eye-catching flash. If a ship notices you, it may send a life raft to collect you. However, planes do not have this option. If you notice the pilot tilting the wing in your direction, they are acknowledging that they have seen you and will send help. Keep all your beacons going until this rescue happens.

Don't lose your life craft! This can be used as a shelter while you create a new one on the island. It may also have rations and equipment which can benefit your survival. Once you have a shelter, think about how you'll get fresh water. To create a solar still on land, all you need is a sheet of clear plastic, a hole, some greenery, and a vessel to catch water. Alternatively, you can dig a sea well behind the first dune of the island. This well should collect filtered seawater and rainwater. Test the drinkability with a taste test to see if it's salty or not.

While on land, there is a chance of a wider array of food. Rock pools are the buffet of the sea. Conches, snails, clams, and even seaweed are all edible, containing the necessary minerals and vitamins you need to keep going a little longer. Continue to scavenge along the coast for food and materials to make your survival easier. Trash can become valuable tools and containers when you have nothing.

Mentally

Surviving a shipwreck is just as traumatic as a plane crash and PTSD will develop over time. Anxiety, sleeping disorders, depression, and even anger issues will be rife among survivors trying to eke out one more day of living. Survival is 90% mental fortitude (AWE me, 2015). If you cannot control your mental status while in a survival situation, it will become impossible to survive day by day. You need to want to survive and see yourself with a future.

However, being over-optimistic about your possible survival is just as bad. Having high hopes of being saved immediately will eventually break you down and cause severe depression. You need to convince yourself that rescue is imminent but may take some time. By remaining grounded in reality, you steel yourself against disappointment but still have hope for the future.

As long as you have hope, you have the strength to face each day alone or with other survivors. Learn to lean on others and recognize signs that someone is distressed about the situation. Keep them working, get tasks done, and work together to keep all of you alive long enough to be rescued.

Final Thoughts

A shipwreck is an ordeal that can last for months out on the sea and years in the survivors' heads. A ship going out from under you, or abandoning you, is terrifying. You lose your shelter, your means of food, water, and security. You are at the mercy of not only the ocean, but the elements, ravenous animals, and your own mind.

Fresh water is hard to come by, and as you slowly start to sweat out your meager supply, your mind starts to play tricks on you. Is that an island? A ship coming to rescue you? Or more disappointment? The survivors discussed in this book had unique experiences that they managed to live through. Some came out on top, some were scarred for years afterward, while others are no longer with us.

Yet in that moment, when they were struggling to keep their heads above water and bailing out their sinking ships, they could only think of one thing. They needed to make it out alive. The mad scramble to grab what supplies they could was instinctual because they had been on the water for so many years before the tragedy occurred.

Each survivor managed to make it to the next chapter of their lives because they understood the risks and what was needed to survive while adrift. Exposure, dehydration, and starvation are the three main killers of those who find themselves adrift at sea. The only way to overcome these hurdles is to ensure you're always fully prepared, whether on a cruise, with a friend fishing, or taking your own yacht out to enjoy a day of sailing.

Thankfully, there are rarely large cruise ships going down, but with personal vessels, there are several thousand accidents a year. The main cause is the misuse of alcohol aboard. Alcohol not only clouds judgment but can lead to excessive dehydration, which causes delusions and hallucinations. When boarding a smaller vessel, take note of the condition of the crew, as this will be an indication of how the vessel is managed.

If you're the captain, you're responsible for those on your vessel. Ensure it's structurally sound, up to the various regulations, and has all the correct and in working order equipment aboard. All life craft should be equipped with EPIRB and emergency supplies. Never carry more people than what your life crafts can carry and ensure all passengers have access to life vests.

Survival is guaranteed if you keep your calm and do as instructed, but you can boost your chances by taking the time to develop a ditch bag that contains all the necessities you'll need to survive. Always think of how you'll construct a shelter and gather food and water. Listen and follow any instructions given by the crew of the ship you're on; this is your most sure way of surviving. When on your own, you only have your preparation and equipment to rely on. If you're ill-prepared, your survival will be difficult and even doomed. Always consider how you'll get shelter, water, and food. This will always aid in your survival, regardless of your situation.

If there is anything we can take from these 10 shipwreck survivors, it's that a cool head and preparedness are what will see you through. Having a few survival skills isn't frowned upon, either. These are skills you can develop while on dry land, so don't wait until you're in the water to learn how to swim. Don't sink before any challenges in your life. Rise to the challenge and swim to the top. Even if it's a doggy paddle, keeping yourself afloat may just be enough to keep you surviving whatever ordeal you're going through.

References

Abrams, A. (2019, August 28). *What do hurricane categories actually mean?* Time. https://time.com/4946730/hurricane-categories/

Alouani, N. (2021, June 16). *This man survived a shipwreck and 76 days alone in the ocean.* Medium. https://historyofyesterday.com/this-man-survived-a-shipwreck-and-76-days-alone-in-the-ocean-d22cc32aafia

Arnarson, A. (2017, June 15). *10 nutrients that you can't get from animal foods.* Healthline. https://www.healthline.com/nutrition/10-nutrients-you-cant-get-from-animal-foods

AWE me. (2015, July 16). *How to survive a shipwreck - epic how to* [Video]. YouTube. https://www.youtube.com/watch?v=XT2y1XD_S_k

Baillie, A. (2018, March 11). *How to fix your position at noon from the sun (meridian passage)* [Video]. YouTube. https://www.youtube.com/watch?v=Jt6l90UOTdI

Barns, I. (Writer), & Barns I. (Director) (2005, October 28) Shark survivor (Season 1, Episode 1) [TV series episode]. In J. Smithson, A. Alani (Executive Producers), *I Shouldn't Be Alive.* Darlow Smithson Productions. Retrieved from https://www.youtube.com/watch?v=rE9_kunIigs&list=RDCMUCFrO-dKhooOuTtix5dia2_g&index=13

Biography Online. (2018, February 28). *Sir Ernest Shackleton biography.* https://www.biographyonline.net/adventurers/ernest-shackleton.html

Biography.com Editors. (2022, March 21). *Ernest Shackleton.* Biography. https://www.biography.com/explorer/ernest-shackleton

Bletchly, R. (2013, January 11). *My real Life of Pi: Incredible story of man who survived 76 days at sea on raft and inspired epic film.* Mirror. https://www.mirror.co.uk/news/real-life-stories/life-of-pi-real-life-shipwecked-1528619

Blitz, M. (2014, December 26). *The man on the raft: The story of Poon Lim.* Today I Found Out.http://www.todayifoundout.com/index.php/2014/12/man-raft-story-poon-lim/

Bogdanowicz, O. (2020, December 20). *133 days stranded at sea.* Medium. https://historyofyesterday.com/133-days-stranded-at-sea-f7f5730521b2

Bryant, C. W. (2008, January 29). *How to survive a shipwreck.* How Stuff Works. https://adventure.howstuffworks.com/how-to-survive-a-shipwreck.htm

Bryant, C. W. (2021, April 12). *How long can you survive adrift in the ocean?* HowStuffWorks. https://adventure.howstuffworks.com/survive-at-sea.htm

Cache Valley Prepper. (2017, June 1). *Survival story: 438 days lost at sea… (video).* Survivopedia. https://www.survivopedia.com/svp-interview-lessons-from-drifting-away/

Callahan, S. (2012, March 23). *Experience: I was adrift on a raft in the Atlantic for 76 days.* The Guardian.https://www.theguardian.com/lifeandstyle/2012/mar/23/adrift-in-atlantic-76-days

Cartwright, M. (2021, September 14). *Alexander Selkirk.* World History Encyclopedia. https://www.worldhistory.org/Alexander_Selkirk/

Chirnside, M. (2011). *The "Olympic" class ships: Olympic, Titanic, Britannic.* History Press.

Clancy, C. J. (2021, September 29). *Steven Callahan: The Irish American who survived being lost at sea.* IrishCentral.https://www.irishcentral.com/opinion/others/steven-callahan-lost-at-sea

Conradt, S. (2020, December 30). *The unsinkable Violet Jessop.* Mental Floss. https://www.mentalfloss.com/article/61663/unsinkable-violet-jessop

Crosta, P. (2017, December 5). *Scurvy: Symptoms, causes, treatment, and prevention.* Medical News Today.

Daniello, V. (2014, September 18). *How to prevent your boat from sinking.* Boating Mag. https://www.boatingmag.com/how-to/how-to-prevent-your-boat-sinking/

Dooley, E., & Gunn, J. (1995). The psychological effects of disaster at sea. *British Journal of Psychiatry, 167*(2), 233–237. https://doi.org/10.1192/bjp.167.2.233

Eatough, R. (Writer); & Eatough, R. (Director) (2010, November 17) 76 days adrift (Season 4, Episode 6) [TV series episode]. In V. McGrath (Executive Producer), *I Shouldn't Be Alive.* Darlow Smithson Productions. Retrieved from https://www.youtube.com/watch?v=A2WXawMmwCo

Encyclopedia Titanica. (n.d.). *Violet Constance Jessop.* https://www.encyclopedia-titanica.org/titanic-survivor/violet-constance-jessop.html

Fitzgerald, C. (2021, June 28). *Mess steward Poon Lim survived 133 days lost at sea.* War History Online.https://www.warhistoryonline.com/instant-articles/poon-lim-survived-133-days-lost-at-sea.html?firefox=1&Exc_D_LessThanPoint002_p1=1

Franklin, J. (2015a, November 7). *Lost at sea: The man who vanished for 14 months.* The Guardian.https://www.theguardian.com/world/2015/nov/07/fisherman-lost-at-sea-436-days-book-extract

Franklin, J. (2015b, November 19). *How Jose Salvador Alvarenga survived 438 days adrift on the open seas.* The Sydney Morning Herald. https://www.smh.com.au/lifestyle/how-jose-salvador-alvarenga-survived-438-days-adrift-on-the-open-seas-20151105-gkrnxx.html

Fredrickson, C. K. (2021, June 30). *U.S. Coast Guard releases 2020 boating safety statistics report.* Coast Guard Maritime Commons Blog. https://mariners.coastguard.blog/2021/06/30/u-s-coast-guard-releases-2020-boating-safety-statistics-report/

George, S. (2021, August 25). *Trashman (+1982).* Wreck Site. https://www.wrecksite.eu/wreck.aspx?306743

Grant, R. (2014, September 3). *How are cruise ships powered.* Cruise Deals Expert. https://cruisedeals.expert/how-are-cruise-ships-powered/

Guinness World Records. (n.d.-a). *Longest time adrift at sea.* https://www.guinnessworldrecords.com/world-records/longest-time-adrift-at-sea

Guinness World Records. (n.d.-b). *Longest time adrift at sea - alone.*https://www.guinnessworldrecords.com/world-records/longest-time-adrift-at-sea-alone

History vs Hollywood. (n.d.). *Adrift (2018).*https://www.historyvshollywood.com/reelfaces/adrift/

How to Survive. (2019, December 23). *How to survive being lost in the ocean, according to science* [Video]. YouTube.https://www.youtube.com/watch?v=z9Oy7HKOcu8

Hunter-Bailey, L. (2021, January 12). *The woman who survived the Titanic, the Olympic, and the Britan-*

nic. Medium.https://historyofyesterday.com/the-woman-who-survived-the-titanic-and-two-other-ship-disasters-b87d89bc52a1

International Chamber of Shipping. (n.d.) . *The principal regulations governing maritime safety.*https://www.ics-shipping.org/shipping-fact/safety-and-regulation-the-principal-regulations-governing-maritime-safety/

IrishCentral Staff. (2021, April 22). *The daughter of Irish immigrants who survived the Titanic, Britannic, and Olympic disasters.* IrishCentral. https://www.irishcentral.com/roots/history/violet-jessop-disasters

Janko. (2019, September 25). *Survival at sea: The thrilling story of Steve Callahan.* SailingEurope. https://www.sailingeurope.com/blog/survival-at-sea-the-thrilling-story-of-steve-callahan

Jet Dock. (n.d.). *What to do when your boat is sinking.* https://www.jetdock.com/knowledge-center/what-to-do-when-your-boat-is-sinking.asp

Keith. (2021, June 29). *Violet Jessop: Irish-Argentine traveller & Titanic survivor.* Nomad Flag. https://nomadflag.com/violet-jessop/

Konrad, T. (2010, July 11). *10 things to remember if you're shipwrecked.* gCaptain. https://gcaptain.com/10-things-to-remember-if-youre-shipwrecked/

Lamoureux, A. (2018, July 23). *The incredible life of Alexander Selkirk, the real Robinson Crusoe.* All That's Interesting. https://allthatsinteresting.com/alexander-selkirk

Lennon, T. (2015, April 5). *Sole survivor of the sinking of the Benlomond in WWII, Poon Lim, set a record for 133 days adrift at sea.* The Daily Telegraph. https://www.dailytelegraph.com.au/news/sole-survivor-of-the-sinking-of-the-benlomond-in-wwii-poon-lim-set-a-record-for-133-days-adrift-at-sea/ news-story/9c63348c427621182e17bcc3c2ddbe1a8

Linder, J. (2015, November 22). *Man's incredible survival story: One year free-floating across the Pacific.* New York Post. https://nypost.com/2015/11/22/how-a-fisherman-survived-more-than-a-year-lost-at-sea/

Macatee, R. (2019, August 1). *The woman whose survival inspired a Shark Week movie lived a truly incredible life.* Distractify. https://www.distractify.com/p/how-did-deborah-scaling-kiley-die

Mahoney, K. (2019, September 27). *Violet Jessop: The world's most unsinkable woman.* Museum Hack. https://museumhack.com/violet-jessop/

Maloney, A. (2019, August 13). *"The sharks got him, then killed the others": Shipwreck survivor recounts five days stranded at sea.* News.com.au. https://www.news.com.au/travel/travel-updates/incidents/the-sharks-got-him-then-killed-the-others-shipwreck-survivor-recounts-five-days-stranded-at-sea/news-story/c60f679af78d537be9e1e09114c26a92

Maloney, A. (2021, July 13). *Drunk captain crashed our boat then sharks ate my friends one by one as I was stranded in bloody sea for 5 days.* The Sun. https://www.thesun.co.uk/news/9671851/shark-week-attack-survivor-ate-friends/

Marriott, L. (1997). *Titanic.* PRC Publishing Ltd.

McCarthy, P. (2015, October 31). *Survival at sea.* Recoil Offgrid. https://www.offgridweb.com/survival/survival-at-sea/

McCarthy, P. (2016, February 14). *The incredible survival story of Alexander Selkirk.* Recoil Offgrid.https://www.offgridweb.com/survival/the-incredible-survival-story-of-alexander-selkirk/

McCarthy, P. (2018, August 19). *The legendary survival story of Sir Ernest Shackleton.* Recoil Offgrid.https://www.offgridweb.com/survival/the-legendary-survival-story-of-sir-ernest-shackleton/

Michelson, R. (2021, July 12). *Though it's Shark Week on TV, sharks are year-round in NC.* Coastal Review.https://coastalreview.org/2021/07/though-its-shark-week-on-tv-sharks-are-year-round-in-nc/

Milford, J. A. (n.d.). *The Olympic-Hawke incident.* Joshua Allen Milford-Titanic. https://www.jmilford-titanic.com/2013/11/september-20-1911-olympic-hawke-accident.html

Mulvaney, K. (2022, March 9). *The stunning survival story of Ernest Shackleton and his endurance crew.* History.https://www.history.com/news/shackleton-endurance-survival

Myall, A. (2021, August 20). *Great survival stories: Shipwreck, sharks, and Deborah Kiley.* Explorersweb.https://explorersweb.com/great-survival-stories-shipwreck-sharks-and-deborah-kiley/

Neaves, S. W. (n.d.). *The storm before the calm.* UTA.https://www.uta.edu/utamagazine/archive-issues/winter_2002/features/storm.html

Nix, E. (2018, August 22). *5 maritime disasters you might not know about.* History. https://www.history.com/news/5-maritime-disasters-you-might-not-know-about

NPR Staff. (2013, February 21). *Real-Life shipwreck survivor helped "Life of Pi" get lost at sea.* NPR.https://www.npr.org/2013/02/24/172632549/real-life-shipwreck-survivor-helped-life-of-pi-get-lost-at-sea

Porter, K. W. (1932). The cruise of the Forester: Some new sidelights on the Astoria Enterprise. *The Washington Historical Quarterly, 23*(4), 261–285. https://www.jstor.org/stable/23908799

Price, M. (2021, July 22). *Sightings of multiple dorsal fins off North Carolina beach trigger shark fears.* The News & Observer. https://www.newsobserver.com/news/state/north-carolina/article252950443.html

Rosenberg, L. (2019, July 31). *The Shark Week flick "Capsized: Blood in the Water" is based on terrifying true events.* Distractify. https://www.distractify.com/p/is-capsized-blood-in-the-water-a-true-story

Rowan, L. (n.d.). *The man who survived 133 days adrift at sea on a wooden raft, WW2.* History Daily. https://historydaily.org/poon li m-survival-story

Royal Caribbean Cruises. (n.d.). *What is a muster drill (safety briefing) on a Royal Caribbean cruise ship?*https://www.royalcaribbean.com/faq/questions/muster-drill-onboard-safety

Royal National Lifeboat Institution. (2019, April 4). *Adrift: Tami Oldham Ashcraft's tale of surviving a hurricane at sea.* https://rnli.org/magazine/magazine-featured-list/2019/april/ad rift-tami-oldham-ashcraft

Salabert, S. (2018, June 7). *The real survival story behind "Adrift."* Outside Online. https://www.outsideonline.com/culture/books-m edia/tami-ashcraft-surviving-41-days-adrift-shailene-woodley/

Santa Barbara Historical Museum. (n.d.). *Captain Jūkichi: The First Japanese immigrant in Santa Barbara.*https://www.sbhistorical.org/captain-jukichi-the-first-japanese-immigrant-in-santa-barbara/

Selcraig, B. (2005, July). *The real Robinson Crusoe.* Smithsonian Magazine. https://www.smithsonianmag.com/history/the-real-robi nson-crusoe-74877644/

Serena, K. (2019, February 15). *The incredible story of José Salvador Alvarenga — who survived 438 days adrift in the pacific.* All That's Interesting. https://allthatsinteresting.com/jose-alvarenga

Serena, K. (2020, October 7). *The true story of "Adrift" and Tami Oldham Ashcraft's survival at sea.* All That's Interesting. https://all thatsinteresting.com/tami-oldham-ashcraft

Shears, R. (2014, February 5). *"What happened to Pinata?" Family of castaway companion demand answers.* Mail Online.

https://www.dailymail.co.uk/news/article-2552160/There-things-dont-match-Castaways-boss-says-doubts-story-13-months-adrift-sea.html

Shimbun, C. (2013, November 1). *Events mark 1800s castaways who were first Japanese in U.S.* The Japan Times. https://www.japantimes.co.jp/news/2013/11/01/national/events-mark-1800s-castaways-who-were-first-japanese-in-u-s/

Smith, O. (2021, June 8). *Deborah Kiley cause of death and life story: Everything you need to know.* Tuko.https://www.tuko.co.ke/414661-deborah-kiley-death-life-story-everything-know.html

Souerbry, R. (2019, November 5). *The real-life story behind the shipwreck movie "Adrift" is almost unbelievable.* Ranker.https://www.ranker.com/list/tami-oldham-ashcraft-survival-story/rachel-souerbry

Spicer, A. (2019, August 20). *What toll does being stranded at sea take on your health?* News4Jax. https://www.news4jax.com/news/2019/08/20/what-toll-does-being-stranded-at-sea-take-on-your-health/

Spilman, R. (2013, December 14). *Poon Lim, surviving a record 133 days at sea — "I hope no one will ever have to break it."* Old Salt Blog. http://www.oldsaltblog.com/2013/12/poon-lim-surviving-a-record-133-days-at-sea-i-hope-no-one-will-ever-have-to-break-it/

The Law Center. (2020, August 24). *Common types and causes of boating accident injuries and what to do if you are suffering from one.* https://thelawcenter.com/personal-injury/articles/common-types-and-causes-of-boating-accident-injuries/

Tikkanen, A. (2017, May 1). *Olympic.* Encyclopædia Britannica. https://www.britannica.com/topic/Olympic

Tikkanen, A. (2021, November 19). *Britannic.* Encyclopædia Britannica. https://www.britannica.com/topic/Britannic

Tikkanen, A. (2022, March 11). *Titanic.* Encyclopædia Britannica. https://www.britannica.com/topic/Titanic/Aftermath-and-investigation

Titor, J. (n.d.). *This man was adrift in the atlantic for 76 days – he survived by using a solar still.* History Daily. https://historydaily.org/story-of-steven-callahan

Travelers Risk Control. (n.d.-a). *How to help prevent your boat from sinking.* Travelers. https://www.travelers.com/resources/boating/how-to-help-prevent-your-boat-from-sinking

Travelers Risk Control. (n.d.-b). *What should I do if my boat is damaged?* Travelers. https://www.travelers.com/resources/boating/what-should-i-do-if-my-boat-is-damaged

Vance, J. E. (2019). *History of ships.* Encyclopædia Britannica. https://www.britannica.com/technology/ship/Passenger-liners-in-the-20th-century

Vonnow, B. (2018, May 19). *Naval disaster: What happened to The Britannic, how did Titanic's sister ship sink and how many people died in the disaster?* The Sun. https://www.thesun.co.uk/news/2297013/britannic-titanic-sister-ship-sink/

Warner, K. (2018, June). *41 days stranded at sea: The harrowing, heartbreaking real-life story behind new movie adrift.* People. https://people.com/movies/41-days-stranded-at-sea-the-harrowing-heartbreaking-real-life-story-behind-new-movie-adrift/

Weird History. (2019, February 3). *Steve Callahan: Survived being adrift at sea for 76 days* [Video]. YouTube. https://www.youtube.com/watch?v=QZ3rylyhoyU

Wilderness Aware Rafting. (2013, January 21). *Story of sea survival: Poon Lim.* https://www.inaraft.com/blog/a-remarkable-story-of-survival-at-sea-poon-lim/

Williams, I. (2018, October 5). *Shackleton's scurvy - or its absence.* Isobel Williams. https://isobelpwilliams.com/2018/10/05/shackletons-scurvy-or-its-absence/

Also by Oliver Matin Cass

SPECIAL BONUS!

Want this Bonus Book for Free?

Get this book for free and also have access to all my new books by joining The Fanbase!

Scan with
your camera
to join!

Made in United States
Orlando, FL
14 January 2025

57263475R00067